T0135650

Studien zur Mustererkennung

herausgegeben von:

Prof. Dr.-Ing. Heinrich Niemann
PD Dr.-Ing. Elmar Nöth

Bibliografische Information der Deutschen Bibliothek

Die Deutsche Bibliothek verzeichnet diese Publikation in der Deutschen Nationalbibliografie; detaillierte bibliografische Daten sind im Internet über http://dnb.ddb.de abrufbar.

ISBN 978-3-8325-1588-1
ISSN 1617-0695

Logos Verlag Berlin
Comeniushof
Gubener Str. 47
10243 Berlin
Tel.: +49 030 42 85 10 90
Fax: +49 030 42 85 10 92
INTERNET: http://www.logos-verlag.de

Appearance-Based Statistical Object Recognition Including Color and Context Modeling

Der Technischen Fakultät der
Universität Erlangen–Nürnberg

zur Erlangung des Grades

DOKTOR–INGENIEUR

vorgelegt von

Marcin Grzegorzek

Erlangen — 2007

Tag der Einreichung:	27.11.2006
Tag der Promotion:	03.04.2007
Dekan:	Prof. Dr. A. Leipertz
Berichterstatter:	Prof. Dr. H. Niemann
	Prof. Dr. R. Tadeusiewicz

Acknowledgments

The work at hand summarizes the main results of my research activities at the Institute of Pattern Recognition in the University of Erlangen-Nuremberg where I have worked as a Research Assistant within the Graduate Research Center for 3D Image Analysis and Synthesis from December 2002 until June 2006. In this point, I would like to express my gratitude to all people who have contributed to this work in any kind.

First of all, I thank my supervisor Prof. Dr. Heinrich Niemann for supporting me by providing necessary incentives, constructive critical questions, encouraging affirmations in due time, as well as for his review of this dissertation and any other suggestion and help. I would like to thank also Prof. Dr. Ryszard Tadeusiewicz who agreed to review this work and to attend my PhD oral exam. Subsequently, I am very thankful to Prof. Dr. Joachim Denzler who was very actively leading me in the first months of my research work. It helped me to define my concrete tasks and scientific direction within the broad field of object recognition. Furthermore, I would like to thank Prof. Dr. Joachim Hornegger for giving me the opportunity to complete the dissertation by extending my scholarship, Prof. Dr. Günther Greiner for a very smooth cooperation within the Graduate Research Center for 3D Image Analysis and Synthesis, and Prof. Dr. Dietrich Paulus for the invitation to the Institute of Pattern Recognition in the summer term 2002.

The system presented in this work has been developed based on the approach introduced by Dr. Michael Reinhold who has spent many hours introducing me into the area of statistical object recognition at the beginning of my research work. In this point, I would like to thank him very much for this great help. I am also very grateful to my girlfriend Marianna Buckan, William E. Minsker, and Dr. Florian Vogt for proofreading. Additionally, I thank Ingo Scholz for our joint research activities and any kind of his help during my work at the Institute of Pattern Recognition.

Finally, I would like to thank my parents Jerzy and Maria Grzegorzek, as well as my brother Dr. Tomasz Grzegorzek for supporting and motivating me during my whole education period.

Marcin Grzegorzek

Contents

Chapter 1

Introduction

Any meaningful human activity requires *perception*. Under perception realization, evaluation, and interpretation of sensory impressions is understood. It allows the human to acquire knowledge about the environment, to react to it, and finally to influence it. There is no reason in principle why perception could not be simulated by some other matter, or instance, a digital computer [Nie90]. The aim of the simulation is not the exact modeling of the human brain activities, but the obtainment of similar perception results. Research activities concerned with the mathematical and technical aspects of perception are the field of *pattern recognition*. One of the most important perceptual abilities is vision. The processing of visual impressions is the task of *image analysis*. The main problem of image analysis is the recognition, evaluation, and interpretation of known patterns or objects in images.

1.1 Fundamental Concept of Object Recognition

One of the most fundamental problems of *computer vision* is the recognition of objects in digital images. The term *object recognition* comprehends both, *classification* and *localization* of objects.

For the problem of object classification, the recognition system should answer the following question:

Which objects appear in the image?

This means that the system determines the classes of objects, which appear in the image, from the set of known object classes $\Omega = \{\Omega_1, \Omega_2, \ldots, \Omega_\kappa, \ldots, \Omega_{N_\Omega}\}$. Generally, the number of objects

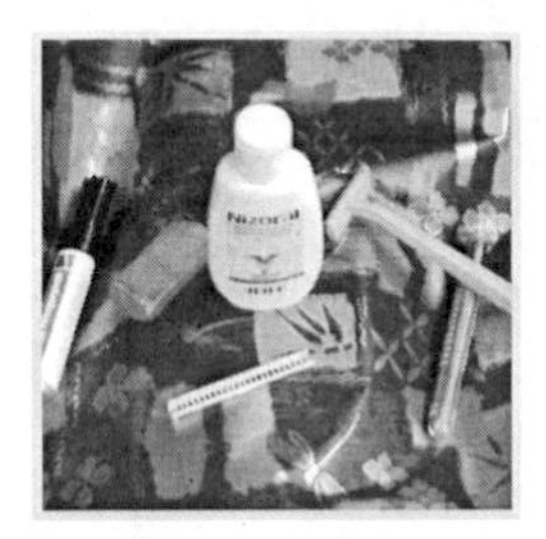

Figure 1.1: Complex scenes with heterogeneous background, where it is difficult to distinguish the objects from the background. The objects can be partly occluded and their number is unknown.

in a scene and their classes is unknown, and any object class can appear many times (see Figure 1.1). Therefore, it is necessary to find out the number of objects in the image first.

In the case of object localization, the recognition system has to find the answer to the question:

In which poses known objects are placed in the image?

The pose of an object in the image is defined with a translation vector $\boldsymbol{t} = (t_x, t_y, t_z)^{\mathrm{T}}$ and three rotation angles (ϕ_x, ϕ_y, and ϕ_z) around the axes of the Cartesian coordinate system. In the present work these three rotation angles are regarded as a rotation vector[1] $\boldsymbol{\phi} = (\phi_x, \phi_y, \phi_z)^{\mathrm{T}}$. The origin of the Cartesian coordinate system is placed in the symmetry center of the image, the x- and y-axes lie in the image plane, and the z-axis is orthographic to the image plane. This object pose definition is illustrated in Figure 1.2. If the object pose is defined with internal transformation parameters $\boldsymbol{t}_{\text{int}} = (t_x, t_y)^{\mathrm{T}}$ and $\phi_{\text{int}} = \phi_z$, the object size and appearance do not change in the image. This two-dimensional case can be seen in the first row of Figure 1.2. More interesting situations occur if the object is transformed with external pose parameters $t_{\text{ext}} = t_z$ and $\boldsymbol{\phi}_{\text{ext}} = (\phi_x, \phi_y)^{\mathrm{T}}$. In the second row of Figure 1.2 the object changes not only the size, but also the appearance for some points of view. The goal of object localization is to determine both, the internal and the external object pose parameters in a real environment image.

Generally, the problem of object recognition, i. e., classification and localization can be expressed with the following question:

Which objects and in which poses appear in the image?

[1]The goal of the object localization is to determine the pose parameters and not to transform any 3D points. Thus, the term rotation vector instead of rotation matrix is used in the present work.

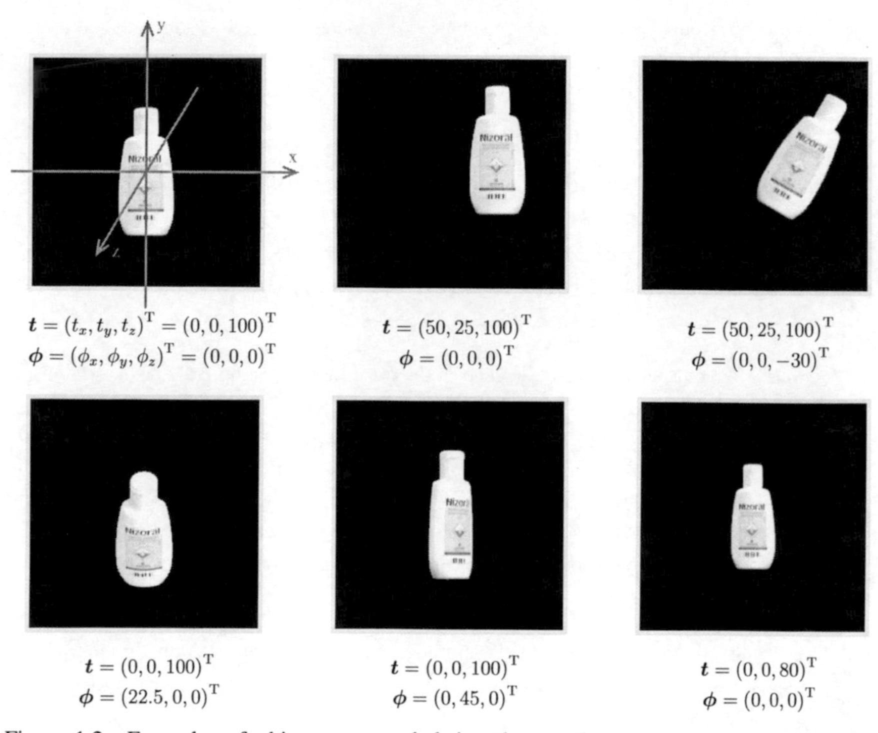

Figure 1.2: Examples of object poses and their values. The components of the internal translation vector $\boldsymbol{t}_{\text{int}} = (t_x, t_y)^{\mathrm{T}}$ are given in pixels, the components of the rotation vector $\boldsymbol{\phi} = (\phi_x, \phi_y, \phi_z)^{\mathrm{T}}$ in degrees [°], and the external translation (scaling) $t_{\text{ext}} = t_z$ in percent [%] of a reference object size (top left).

There are two main approaches for object recognition. First, there exist *shape-based* approaches that perform a segmentation and use geometric features like lines or corners [Hor96, Ker03, Che04, Lat05]. These methods suffer from segmentation errors and have problems to deal with objects, which have no distinct edges. Therefore, many authors, e. g., [Mur95, Pös99, Rei04], prefer a second method, the *appearance-based* approach. Here, the features are directly computed from the pixel values without a previous segmentation step. A detailed description of both methods is given in Chapter 3.

As can be seen in Figure 1.1 it is even for humans sometimes difficult to distinguish objects from background. The objects can be placed on a heterogeneous background and they can be partly occluded. Nevertheless, a robust object recognition system should be able to handle problems like:

- *Heterogeneous Background*
  In real environments objects are placed in images on a heterogeneous background (see Figure 1.1). Often other objects appear in the neighborhood of an object. This can change the object appearance because of shadow or illumination reflection. Nevertheless, the object recognition system must be able to find the objects in such an environment.

- *Occlusions*
  Very often objects which are searched in the image are not completely visible (see Figure 1.1). They can be occluded by other objects or obstacles. In such cases the system should also correctly classify and localize these objects.

- *Illumination Changes*
  It cannot be assumed that the illumination is constant in a real environment. The object recognition system should work independent on illumination changes.

- *Multi-Object Scenes*
  Generally, the number of objects in an image is unknown. The additional problem for the object classification and localization system is to determine the number of objects which appear in the image.

Figure 1.3 shows a scheme of a basic object recognition system, which works in two phases, namely training and recognition phase. For the training phase images of all objects from different viewpoints are taken. The images are then preprocessed and feature vectors are computed. After that for all object classes $\Omega = \{\Omega_1, \Omega_2, \ldots, \Omega_\kappa, \ldots, \Omega_{N_\Omega}\}$ which are considered in the recognition task, the system creates object models $\mathcal{M} = \{\mathcal{M}_1, \mathcal{M}_2, \ldots, \mathcal{M}_\kappa, \ldots, \mathcal{M}_{N_\Omega}\}$. In the recognition phase an image from a real environment is taken, preprocessed, and feature vectors are computed with the same method as in the training phase. Then the algorithm for object classification and localization, which compares the extracted feature vectors with the created object models, determines the classes and poses of objects which were found in the image [Nie83]. All these steps are described in more detail later.

For many tasks the classification and localization of objects in images is very useful, sometimes even necessary. Algorithms for automatic computational object recognition can be applied for example to:

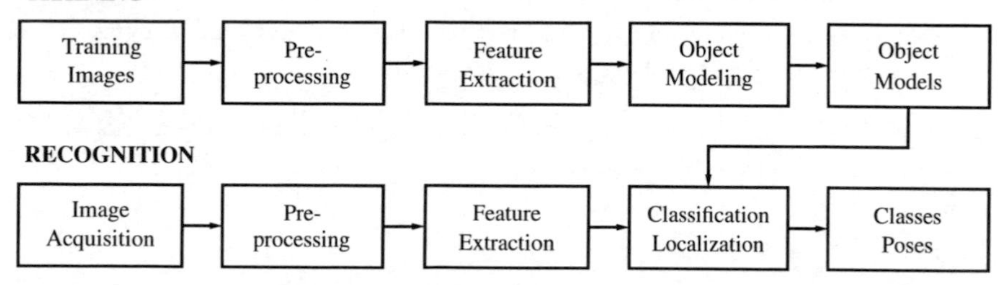

Figure 1.3: General scheme of an object recognition system with the training and recognition phase [Nie83].

- *Face Classification [Gro04, Ter04]*
  Nowadays different systems are applied, which classify persons based on their face appearance. The systems are able to find a particular person in a large database within few seconds.

- *Fingerprint Classification [Zha04, Par05]*
  Nowadays criminology without automatic fingerprint classification could not be imagined.

- *Handwriting Recognition [Cho04, Heu04]*
  There are many tasks, for which it would be useful to read, interpret, and process handwritten text. Think of all the forms which everyone of us fills out every year!

- *Service Robotics [You03, Zob03]*
  Service robots should be able not only to find (classify) a particular object, e. g., in the kitchen, but also to localize it. In order to take an object into the hand, you should exactly know where it is.

- *Medicine [Ben02, Li02]*
  Computers can help doctors to diagnose patients. A large part of medical data exists in the form of digital images. In these images special recognition systems can find pathological areas, and give the doctor and the patient a first warning.

- *Visual Inspection [Kum03, Nga05]*
  Object recognition systems are used for quality inspection of serially produced units at production lines.

- *Automobile Industry [Frü04, Gau05]*
  First experiments with an infrared camera mounted on a driving car were made. The objective was to localize obstacles on the road in the night.

Although for some applications object recognition systems have been already developed, e. g., fingerprint classification, there are still enough areas where such systems would be useful and simplify our life.

1.2 Contribution of the Present Work

This work is a continuation of the PhD thesis from Dr. Michael Reinhold. Dr. Reinhold presented in [Rei04] a system for appearance-based statistical object recognition with local feature vectors. The first goal of the present work is the improvement of the recognition rates for 3D scenes with real heterogeneous background in comparison to [Rei04] . In [Rei04] the system was evaluated based on artificially created test images. This means that the objects were taken on a dark background and then pasted on a heterogeneous background, which was taken separately. It is not taken into account that objects can change their appearance depending on the background in the image acquisition process. In the present work all experiments are performed using test images with real heterogeneous background. Another objective of this work is context modeling in the case of multi-object scenes. In [Rei04] objects were considered independent on each other in multi-objects scenes. However, there are situations where one can expect an occurrence of an object if a particular object or objects were already found in the image.

The following main aspects are new in the present work:

- *Fusion of Multiple Views*
  A new approach for the fusion of multiple views based on a recursive density propagation method is introduced. In contrast to passive algorithms, where the decision about class and pose of an object has to be taken based on one image, here more images are used. The additional images are used to gain more information about the scene and the observed objects. The experimental results show that the fusion improves the recognition rates substantially, especially for difficult conditions like heterogeneous background within real world environments.

- *Fast Training*
  A new approach is proposed, where the image acquisition for the training phase is done with a hand-held camera. The poses of the objects in all training frames are computed

using a structure-from-motion algorithm [Hei04]. The whole learning process is, therefore, independent on environment assumptions, but we have to deal with an additional training inaccuracy.

- *Resolution Level Combination of Wavelet Transformation*
  Feature extraction on three different resolution levels of the wavelet transformation is introduced, and three statistical object models for each object class are created in the training phase. The algorithm for object classification and localization uses a combination of the object models obtained for these different resolution levels, which significantly improves the recognition rates.

- *Color Modeling*
  In [Rei04] object models were created based on gray level images. In the present work it is proposed to use the color information of objects.

- *Multi-Object Scenes with Context Modeling*
  In [Rei04], multi-object scenes were considered without attention to the context dependencies. In the present work, context modeling for multi-object scenes is introduced.

- *3D-REAL-ENV*
  In [Rei04], the system was evaluated based on the DIROKOL image database [Rei01] with artificially created test images. In the present work, experiments are performed using test images with real heterogeneous background. For this reason the image database for 3D object recognition in real world environment (3D-REAL-ENV) with more than 5000 test images with real heterogeneous background was generated. The most algorithms are evaluated using the 3D-REAL-ENV database, which makes their comparison very objective.

1.3 Motivation

In the previous section, new system components in comparison to [Rei04] are described. The present section gives a motivation, why the new components are introduced. Additionally, the choice of the statistical appearance-based approach with local feature vectors is explained.

- *Fusion of Multiple Views*
  There are recognition tasks in which objects cannot be distinguished from some viewpoints, even by humans. Figure 1.4 shows two bottles which look identical for several angles of the external rotation ϕ_y. In such cases a correct classification and localization

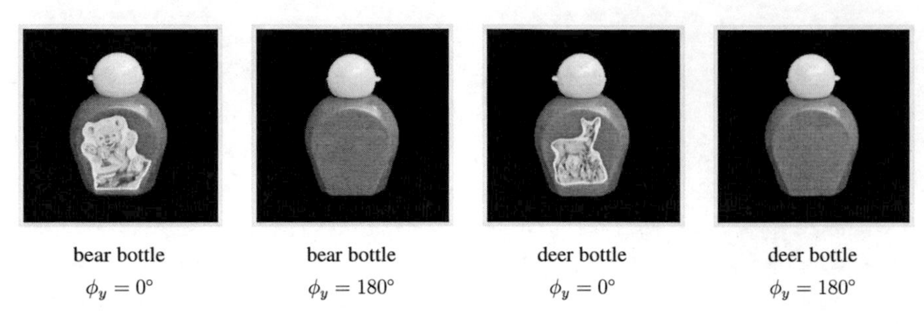

Figure 1.4: Two different objects, which cannot be distinguished from several viewpoints, e. g., $\phi_y = 180°$.

using only one image is not possible. This is the reason why the algorithm for fusion of multiple views based on the condensation algorithm [Isa98] is introduced in the present work. The decision about class and pose of an object is made using more than one image.

- *Fast Training*
  The learning process in most object recognition systems begins with the image acquisition of all possible objects from different viewpoints using a special setup like a turntable with a camera arm. For real tasks of object classification and localization it is much easier to record the objects using a hand-held camera. Think of very big or constantly moving objects like animals! The pose parameters of the training images, which are needed for creating the object models, are computed using a structure-from-motion algorithm [Hei04].

- *Resolution Level Combination of Wavelet Transformation*
  In [Rei04], object feature vectors were computed using only one scale of the wavelet transformation. In the present work, a combination of different resolution levels of the wavelet transformation is used. Thus, the recognition result is obtained on a low resolution, and additionally reviewed on higher resolutions. This repeated result refinement increases the probability to get a correct class and pose of an object in a scene.

- *Color Modeling*
  Many systems for object classification and localization do not use the color information of objects. Think of it, how often humans make use of color information due to distinguish objects! Sometimes two or more objects having totally different colors look similar in gray level images. It makes sense to introduce color modeling to the object recognition system.

- *Multi-Object Scenes with Context Modeling*
  In [Rei04], multi-object scenes were considered without regarding the context dependencies. There are situations in real world environment, in which a context information about a particular scene can help the system to classify and localize objects. In the present work, the context dependencies between objects in multi-object scenes are modeled.

- *3D-REAL-ENV*
  In order to allow an objective evaluation of recognition algorithms which are introduced within the scope of this work, the image database for 3D object recognition in real world environment (3D-REAL-ENV) with more than 5000 test images with real heterogeneous background was generated.

- *Statistical Approach*
  The goal of the object recognition system is the classification and localization of objects within real world environment. In such an environment the object appearance varies and depends on the background, possible occlusions, and illumination changes. The simple matching of object feature vectors which are extracted in the training phase, with the features from the real recognition scene yields very bad results even if there is the same object in the image. The only chance to make a correct classification and localization within real world environment is the statistical modeling of the feature vectors in the learning phase. The object features have to be represented as density functions, e. g., by modeling with the normal distribution. In the present work, the object models are integrated into a statistical framework, and so the system can deal with noise and illumination changes.

- *Appearance-Based Approach*
  For recognition of 3D objects in 2D images, two main approaches in computer vision are known: based on the results of a segmentation (shape-based), or directly on the object appearance (appearance-based). Segmentation operations detect geometric features such as lines or corners. These features as well as the relations between them are used for object recognition [Hor96]. However, the segmentation approaches suffer from two disadvantages: segmentation errors, and loss of information contained in the image caused by the segmentation. In contrast to this, appearance-based approaches avoid these disadvantages. They use the image data, i. e., the pixel values, directly without a previous segmentation step. The simplest method is correlation of an image with an object template. Another method is the eigenspace approach [Grä03]. Thereby a large number of images are encoded by a small number of basis images, so called eigenimages. The object recognition

system which is presented in this work, should be able to work with different kinds of objects, and cannot assume that the object shape is known. This is the reason why the appearance-based way for the development of this system is chosen.

- *Local Feature Vectors*
  There exist object recognition systems which use only one global feature vector for the whole image, e. g., eigenspace approach [Grä03], others describe objects with several local features, e. g., neural networks [Par04]. If only one pixel in the image changes its value because of noise, the only one global feature vector, which describes the object, is affected. Due to this disadvantage, in the present work the objects are represented by several local features for each image.

1.4 Overview

Chapter 2 describes the mathematical instruments which were used for development of the object recognition system. The knowledge of fundamentals of statistics, wavelet transformation, and theory of function approximation is necessary for understanding the statistical object modeling.

Chapter 3 gives an overview over existing algorithms and methods for object recognition in digital images. First, some fundamentals of this area are presented, which is followed by the description of known shape-based and appearance-based approaches.

The training phase of the system is discussed in Chapter 4. After training data collection, statistical object modeling using gray level images, and the statistical modeling using color images, the chapter presents the learning of the context dependencies for multi-object scenes.

In Chapter 5 the recognition phase of the system is described. First, algorithms for object classification and localization in single-object scenes, then, object recognition for multi-object scenes are discussed.

Experiments and results are presented in Chapter 6. First, the experimental environment, the evaluation criteria, and the new 3D-REAL-ENV image database are described. After that, all recognition algorithms shown in Chapter 5 are experimentally evaluated and compared.

After the experimental evaluation of all algorithms for object classification and localization in Chapter 6, the work is closed with a conclusion in Chapter 7.

Chapter 2

Mathematical Background

This chapter describes the mathematical instruments which are used for the development of the object recognition system. In Section 2.1, fundamentals of mathematical statistics are explained, because the object models in the present work are integrated into a statistical framework. The object recognition system computes the feature vectors using the wavelet transformation. Therefore, its theory is presented in Section 2.2. Due to function approximation (Section 2.3), it is possible to come from the discrete pose parameter domain in the training phase to the continuous pose parameter domain in the recognition phase. Section 2.4 summarizes the present chapter.

2.1 Fundamentals of Statistics

Statistics is the science and practice of developing knowledge through the use of empirical data expressed in quantitative form. It is based on statistical theory, which is a branch of applied mathematics. Within statistical theory, randomness and uncertainty are modeled by probability theory. Because one aim of statistics is to produce the "best" information from available data, some authors consider statistics a branch of decision theory [Bam72]. Statistical practice includes the planning, summarizing, and interpreting of observations, allowing for variability and uncertainty. *Mathematical statistics* uses probability theory and other branches of mathematics to study statistics from a purely mathematical standpoint.

A *random variable* is a function $X(\lambda)$ assigning a number x to every outcome λ of an experiment. This number could be the gain in a game of chance, the voltage of a random source, the cost of a random component, a pixel value $(0, 1, \ldots, 255)$ of a digital gray level image, or any other numerical quantity that is of interest in the performance of the experiment. Thus, the random variable is a function whose domain is the set Λ of all possible experiment outcomes λ, and the range a set of numbers [Pap84]. Generally, the range of the function X can be a set of complex numbers. However, in the present section only real random variables $X : \Lambda \rightarrow \mathbb{R}$ are considered, because the "less than or equal" relation is needed for the definition of the distribution function. If the function $X : \Lambda \rightarrow \mathbb{R}$ is continuous (discrete), we speak about a *continuous (discrete) random variable* X. Under a *random vector* a vector is understood, whose components are random variables. The $N_{\boldsymbol{X}}$-dimensional random vector $\boldsymbol{X}$ can be written as

$$\boldsymbol{X} = (X_1, X_2, \ldots, X_{N_{\boldsymbol{X}}})^{\mathrm{T}} \quad . \tag{2.1}$$

Let's introduce the following notation

$$\{X \leq x\} \quad . \tag{2.2}$$

This notation represents a subset of Λ consisting of all experiment outcomes λ such that $X(\lambda) \leq x$. The meaning of

$$\{x_a \leq X \leq x_b\} \tag{2.3}$$

is similar. It represents a subset of Λ consisting of all experiment outcomes λ such that $x_a \leq X(\lambda) \leq x_b$. Finally the notation

$$\{X = x\} \tag{2.4}$$

is a subset of Λ consisting of all experiment outcomes λ such that $X(\lambda) = x$. With $P(\{X \leq x\})$, $P(\{x_a \leq X \leq x_b\})$, and $P(\{X = x\})$ the probabilities of the events (subsets of Λ) defined in (2.2), (2.3), and (2.4) are denoted. Now a formal definition of the real random variable can be introduced.

Definition 2.1 *A real random variable X is a function $X : \Lambda \ni \lambda \rightarrow x \in \mathbb{R}$ assigning a real number $x = X(\lambda)$ to every experiment outcome λ, where the following two conditions have to be fullfiled:*

1. *The set $\{X \leq x\}$ is an event for every x,*

2. *The probabilities of the events $\{X = \infty\}$ and $\{X = -\infty\}$ are equal to zero, i. e.,*

$$P(\{X = \infty\}) = 0 \quad \textit{and} \quad P(\{X = -\infty\}) = 0 \quad . \tag{2.5}$$

The second condition states that, although we allow $X(\lambda)$ to be ∞ or $-\infty$ for some experiment outcomes λ, we demand that these outcomes form an event (subset of Λ), for which the probability is equal to zero.

The elements of the subset of Λ that are contained in the event $\{X \leq x\}$ change as the number x takes various values. The probability $P(\{X \leq x\})$ is, therefore, a number that depends on x. This number is denoted by $F(x)$ and is called the *distribution function* of the random variable X.

Definition 2.2 *The distribution function of the random variable X is the function*

$$F(x) = P(\{X \leq x\}) \tag{2.6}$$

defined for every x from $-\infty$ to ∞.

The distribution function has the following properties:

1. $F(-\infty) = 0$ and $F(\infty) = 1$,
2. If $x_a < x_b$, then $F(x_a) \leq F(x_b)$,
3. If $F(x_0) = 0$, then $F(x) = 0$ for every $x \leq x_0$,
4. $P(\{X > x\}) = 1 - F(x)$,
5. $P(\{x_a < X < x_b\}) = F(x_b) - F(x_a)$.

Definition 2.3 *The derivative*

$$p(x) = \frac{dF(x)}{dx} \tag{2.7}$$

of $F(x)$ is called the density function of the continuous random variable X.

If the random variable X is of discrete type taking the values x_i with probabilities p_i, then

$$p(x) = \sum_i p_i \delta(x - x_i) \quad , \tag{2.8}$$

where $p_i = P(\{X = x_i\})$, and $\delta(x)$ is an impulse function. An example distribution function $F(x)$ and the corresponding density function $p(x)$ for a discrete random variable X is depicted in Figure 2.1. The term $p_i\delta(x - x_i)$ is shown as a vertical arrow at $x = x_i$ with length equal to p_i. The density function of a continuous[1] random variable has the following properties:

[1]The density function of a discrete random variable is not integrable. In this case, its values have to be added to each other, and the integral symbol $\int$ replaced by the sum symbol $\sum$.

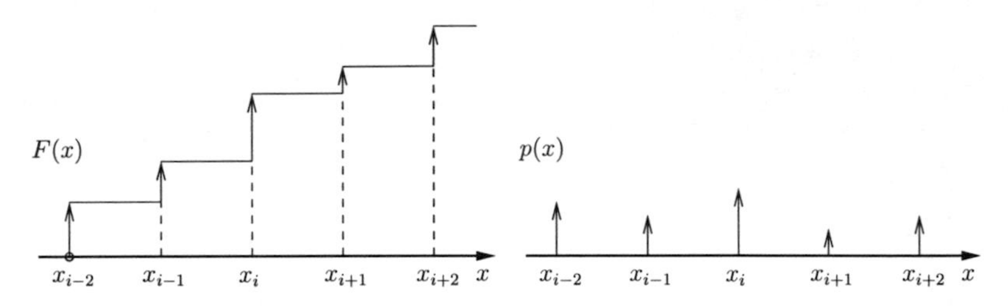

Figure 2.1: An example distribution function $F(x)$ and the corresponding density function $p(x)$ for a discrete random variable X.

1. From the monotonicity of $F(x)$ it follows that $p(x) \geq 0$ for all x,

2. Integrating (2.7) from $-\infty$ to x and using the fact that $F(-\infty) = 0$, we obtain

$$F(x) = \int_{-\infty}^{x} p(\tau) d\tau \quad , \tag{2.9}$$

3. Since $F(\infty) = 1$, the above yields

$$\int_{-\infty}^{\infty} p(x) dx = 1 \quad , \tag{2.10}$$

4. From (2.9) it follows that

$$F(x_b) - F(x_a) = \int_{x_a}^{x_b} p(x) dx \quad , \tag{2.11}$$

5. Hence

$$P(\{x_a < X \leq x_b\}) = \int_{x_a}^{x_b} p(x) dx \quad . \tag{2.12}$$

Definition 2.4 *The expected value or mean μ of a continuous random variable X is by definition the integral*

$$\mu = \int_{-\infty}^{\infty} x p(x) dx \quad . \tag{2.13}$$

In the case of a random vector $\boldsymbol{X} = (X_1, X_2, \ldots, X_{N_{\boldsymbol{X}}})^{\mathrm{T}}$ we speak about a *mean vector*

$$\boldsymbol{\mu} = (\mu_1, \mu_2, \ldots, \mu_{N_{\boldsymbol{X}}})^{\mathrm{T}} \quad . \tag{2.14}$$

Definition 2.5 *The variance σ^2 of a continuous random variable X is by definition the integral*

$$\sigma^2 = \int_{-\infty}^{\infty} (x - \mu)^2 p(x) dx \quad . \tag{2.15}$$

The positive constant σ is called standard deviation of X.

For a random vector $\boldsymbol{X} = (X_1, X_2, \ldots, X_{N_{\boldsymbol{X}}})^{\mathrm{T}}$ a *covariance matrix* is defined as

$$\boldsymbol{\Sigma} = \begin{pmatrix} \sigma_{11}^2 & \ldots & \sigma_{1N_{\boldsymbol{X}}}^2 \\ \vdots & & \vdots \\ \sigma_{N_{\boldsymbol{X}}1}^2 & \ldots & \sigma_{N_{\boldsymbol{X}}N_{\boldsymbol{X}}}^2 \end{pmatrix} \quad . \tag{2.16}$$

For $i \neq j$ σ_{ij}^2 denotes a covariance between the components X_i and X_j, and for $i = j$ a variance of the component X_i. If two components X_i and X_j $(i \neq j)$ of the random vector $\boldsymbol{X}$ are statistically independent, then

$$\sigma_{ij} = \sigma_{ji} = 0 \quad , \tag{2.17}$$

but in general not vice versa. If all components of the random vector $\boldsymbol{X}$ are statistically independent, the covariance matrix is reduced to the diagonal matrix

$$\boldsymbol{\Sigma} = \begin{pmatrix} \sigma_{11}^2 & 0 & \ldots & 0 \\ 0 & \sigma_{22}^2 & 0 & \vdots \\ \vdots & 0 & \ddots & 0 \\ 0 & \ldots & 0 & \sigma_{N_{\boldsymbol{X}}N_{\boldsymbol{X}}}^2 \end{pmatrix} \quad . \tag{2.18}$$

A random variable X is called *normal* or *Gaussian* if its density function is described with the following equation

$$p(x) = \frac{1}{\sigma\sqrt{2\pi}} \exp\left(\frac{(x - \mu)^2}{-2\sigma^2}\right) \quad . \tag{2.19}$$

Some plots of one-dimensional Gaussian density functions for given mean values μ and standard deviations σ can be seen in Figure 2.2. A random vector $\boldsymbol{X} = (X_1, X_2, \ldots, X_{N_{\boldsymbol{X}}})^{\mathrm{T}}$ is called

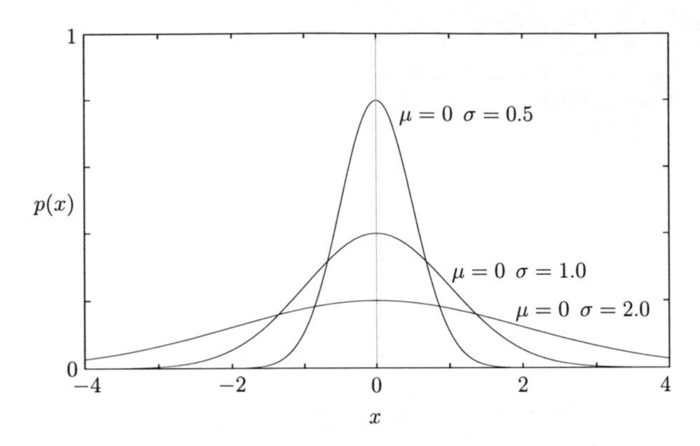

Figure 2.2: Some plots of one-dimensional Gaussian density functions for given means μ and standard deviations σ.

normal or Gaussian if its density function has the following definition

$$p(\boldsymbol{x}) = \frac{1}{\sqrt{\det(2\pi\boldsymbol{\Sigma})}} \exp\left(\frac{(\boldsymbol{x}-\boldsymbol{\mu})^{\mathrm{T}}\boldsymbol{\Sigma}^{-1}(\boldsymbol{x}-\boldsymbol{\mu})}{-2}\right) \quad , \tag{2.20}$$

where $\boldsymbol{x} = (x_1, x_2, \ldots, x_{N_{\boldsymbol{X}}})^{\mathrm{T}}$, and $x_i = X_i(\lambda)$. An example plot of a two-dimensional Gaussian density function is depicted in Figure 2.3. If all components X_i of the random vector $\boldsymbol{X}$ are statistically independent, then (2.20) can be written as a product of $N_{\boldsymbol{X}}$ one-dimensional normal density functions (2.19)

$$p(\boldsymbol{x}) = \prod_{i=1}^{N_{\boldsymbol{X}}} p(x_i) = \prod_{i=1}^{N_{\boldsymbol{X}}} \left(\frac{1}{\sigma_{ii}\sqrt{2\pi}} \exp\left(\frac{(x_i-\mu_i)^2}{-2\sigma_{ii}^2}\right)\right) \quad . \tag{2.21}$$

A random variable X is called *uniform* between x_a and x_b if its density is constant in the interval $[x_a, x_b]$ $(x_a < x_b)$ and zero elsewhere

$$p(x) = \begin{cases} \frac{1}{x_b - x_a}, & \text{if} \quad x_a \leq x \leq x_b \\ 0, & \text{otherwise} \end{cases} \quad . \tag{2.22}$$

An example distribution and density functions for a uniform random variable are depicted in Figure 2.4.

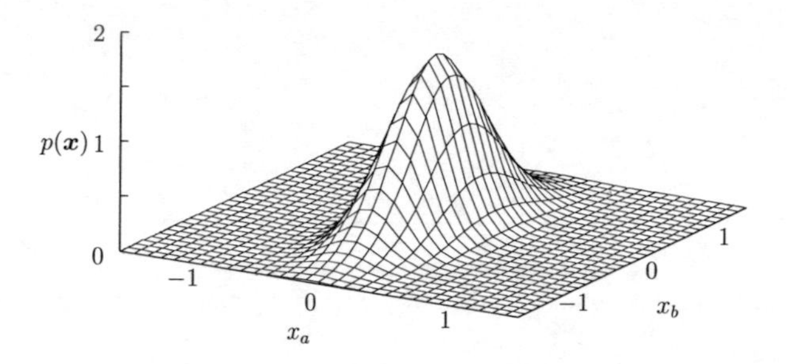

Figure 2.3: An example of a two-dimensional Gaussian density function with $\sigma_{11} = 0.2$, $\sigma_{22} = 0.5$, and $\sigma_{12} = \sigma_{21} = 0.1$.

For two random variables, X and Y a *conditional probability density* $p(x|y)$ which denotes the probability density of the event $\{X = x\}$ (2.4), if the occurrence of the event $\{Y = y\}$ is given, can be computed as follows

$$p(x|y) = \frac{p(x)p(y|x)}{p(y)} \quad . \tag{2.23}$$

A *join probability density p(x, y)* for the random variables X and Y, i. e., the probability density to observe both events $\{X = x\}$ and $\{Y = y\}$ is given by

$$p(x, y) = p(y)p(x|y) = p(x)p(y|x) \quad . \tag{2.24}$$

If random variables X and Y are statistically independent on each other, then the conditional probability (2.23) can be simplified to

$$p(x|y) = p(x) \quad \text{and} \quad p(y|x) = p(y) \quad . \tag{2.25}$$

In this case from (2.24) and (2.25) it follows

$$p(x, y) = p(x)p(y) \quad . \tag{2.26}$$

If the conditional probability density $p(x|y)$ is given and continuous random variables X and Y are statistically dependent on each other, then the probability density $p(x)$ can be computed with

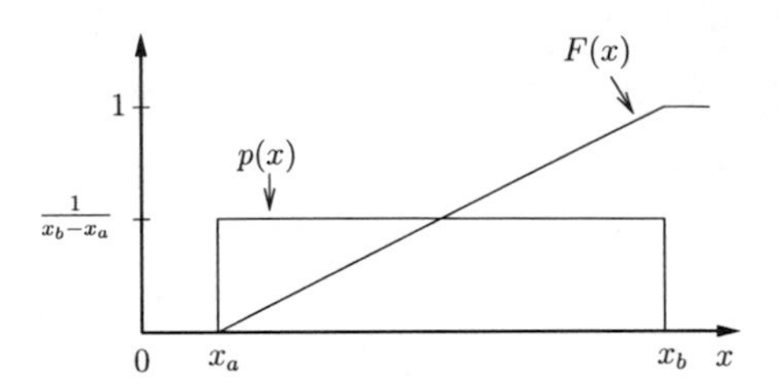

Figure 2.4: An example distribution and density functions for a uniform random variable.

the so-called marginalization rule

$$p(x) = \int_{-\infty}^{\infty} p(x,y)dy = \int_{-\infty}^{\infty} p(y)p(x|y)dy \quad . \tag{2.27}$$

It is sometimes necessary to determine the value $\hat{y}$ of a random variable Y for a given density $p(x)$, which maximizes the probability density $p(y|x)$. But often only the density $p(x|y)$ is known. In this case, a so-called *maximum a-posteriori estimation (MAP)* is applied (e. g., [Nie90, Web02])

$$\hat{y} = \underset{y}{\operatorname{argmax}}\, p(y|x) = \underset{y}{\operatorname{argmax}} \frac{p(y)p(x|y)}{p(x)} = \underset{y}{\operatorname{argmax}}\, p(y)p(x|y) \quad . \tag{2.28}$$

Instead of maximizing the conditional probability density $p(y|x)$, the product $p(y)p(x|y)$ is maximized according to the value y of the random variable Y.

2.2 Wavelet Transformation

Everywhere around us are signals that can be analyzed. For example, there are seismic tremors, human speech, engine vibrations, medical images, financial data, music, and many other types of signals. *Wavelet analysis* is a new and promising set of tools and techniques for analyzing these signals. In the present work, the two-dimensional discrete wavelet transformation is applied to extraction of object feature vectors for object modeling. In Section 2.2.1, the *Fourier transformation* of continuous one-dimensional functions is defined, and its main difference in comparison to the *wavelet transformation* shortly commented. The theory of the wavelet transformation for continuous one-dimensional signals is presented in Section 2.2.2. As images are two-dimensional discrete signals, Section 2.2.3 introduces the wavelet analysis of discrete functions with the extension to the two-dimensional case.

2.2.1 From Fourier Analysis to Wavelet Transformation

For some applications it is easier to use a signal in the frequency domain instead of spatial domain. The basic idea of using the frequency domain is founded on the mathematical result that an arbitrary 2π-periodic function[2] $f : \mathbb{R} \to \mathbb{R}$ can be represented by a *Fourier series*

$$f(x) = \frac{a_0}{2} + \sum_{\omega \geq 1} (a_\omega \cos(\omega x) + b_\omega \sin(\omega x)) \quad , \tag{2.29}$$

i. e., a superposition of weighted sine and cosine terms [Pau03]. The scalar weights a_ω and b_ω are called the *Fourier coefficients* and can be computed as follows

$$a_\omega = \frac{1}{\pi} \int_{-\pi}^{\pi} f(x) \cos(\omega x) dx \quad , \tag{2.30}$$

and

$$b_\omega = \frac{1}{\pi} \int_{-\pi}^{\pi} f(x) \sin(\omega x) dx \quad . \tag{2.31}$$

The Fourier series (2.29) can also be written in the complex form

$$f(x) = \frac{1}{2\pi} \sum_{\omega=-\infty}^{\infty} c_\omega \exp(j\omega x) \quad , \tag{2.32}$$

where

$$c_\omega = \begin{cases} \pi(a_\omega - jb_\omega), & \text{if } \omega \geq 0 \\ \pi(a_{|\omega|} + jb_{|\omega|}), & \text{otherwise} \end{cases} \quad , \tag{2.33}$$

and j is the imaginary unit ($j^2 = -1$). The formula for computing the weights c_ω is shown to be

$$c_\omega = \int_{-\pi}^{\pi} f(x) \exp(-j\omega x) dx \quad . \tag{2.34}$$

[2] $f(x) = f(x + 2\pi)$

Let us assume that the interval of periodicity for the function $f(x)$ is infinite. The sum of (2.32) will become an integral and the coefficients c_ω will be continuous weight functions $c(\omega)$, i. e.,

$$c(\omega) = \int_{-\infty}^{\infty} f(x) \exp(-j\omega x) dx \quad . \tag{2.35}$$

The weight function $c(\omega)$ is called the *Fourier transform* of the function $f(x)$.

Definition 2.6 *The continuous Fourier transform $F(\omega)$ defined on the frequency domain of the function $f(x)$ defined on the spatial domain is defined by (e. g., [Cas96, Pau03])*

$$F(\omega) = \int_{-\infty}^{\infty} f(x) \exp(-j\omega x) dx = \mathrm{FT}\{f\} \quad . \tag{2.36}$$

The inverse of the continuous Fourier transform is given with the following equation

$$f(x) = \frac{1}{2\pi} \int_{-\infty}^{\infty} F(\omega) \exp(j\omega x) d\omega \quad . \tag{2.37}$$

The Fourier transform of a function represents the amplitude of each frequency found in the signal. It means that each signal defined on the spatial domain can be transformed into the frequency domain by (2.36), processed in it, and back transformed into the spatial domain by (2.37). But the Fourier transformation has also some disadvantages. Its basis function is of a trigonometric character (wave)

$$\exp(-jx) = \cos x - j \sin x \quad . \tag{2.38}$$

It does not asymptotically converge to zero for $x \to -\infty$ and $x \to \infty$. It means that after the Fourier transformation, a local signal[3] in the spatial domain takes a large spectrum in the frequency domain. For example, a thin edge in a digital gray level image (two-dimensional discrete signal) spreads on a large frequency spectrum after the Fourier transformation. Thus, the information that it was an edge gets lost with the Fourier transformation.

The wavelet transformation is an approach which was introduced to compensate this disadvantage. Instead of waves as basis functions it uses so-called *wavelets*. A graph of example wave and wavelet is depicted in Figure 2.5. One can see immediately what the difference is.

[3] A signal, which values are not equal to zero only for a small interval $x \in [x_1, x_2]$.

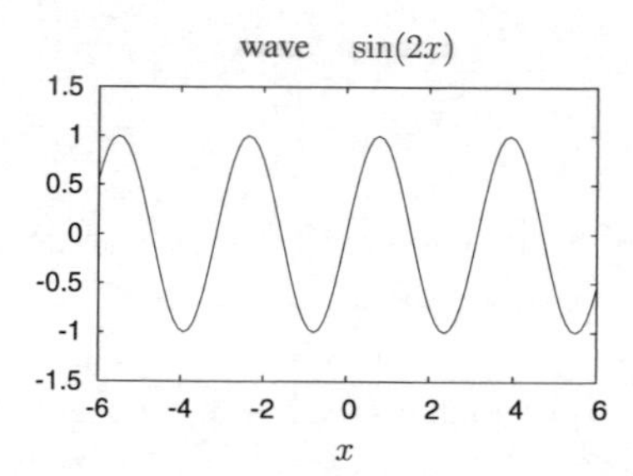

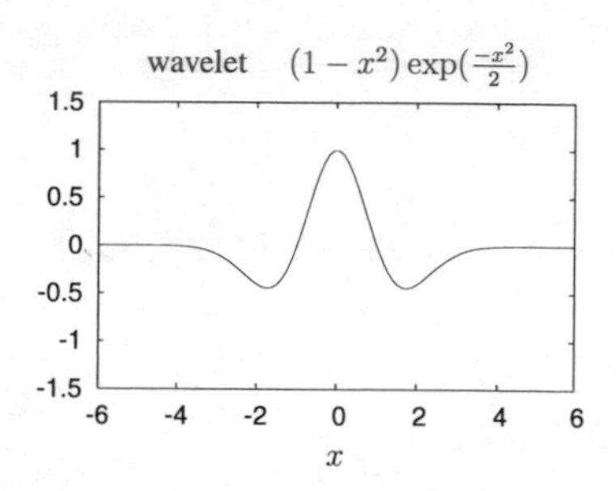

Figure 2.5: Left: an example basis function for the Fourier transformation (wave). Right: an example basis function for the wavelet transformation (wavelet).

The wavelet converges to zero, whereas the wave is not converging for $x \to -\infty$ and $x \to \infty$. This property of the wavelet basis functions result in better transformation of local signals. The locality information of a signal does not get lost in the case of wavelet transformation.

2.2.2 Continuous Wavelet Transformation

Under *continuous wavelet transformation*, a transformation of continuous functions is understood, but the result (*wavelet transform*) can be continuous or discrete. Only functions which are *square integrable* on the set of real numbers $\mathbb{R}$, can be transformed with the continuous wavelet transformation (e. g., [Dau90, Chu92]).

Definition 2.7 *A real function $f(x)$ is square integrable on $\mathbb{R}$ if the integral over $\mathbb{R}$ of the square of its absolute value is finite*

$$\int_{-\infty}^{\infty} |f(x)|^2 dx < \infty \quad . \tag{2.39}$$

The set of all square integrable functions on $\mathbb{R}$ is denoted by $L^2(\mathbb{R})$.

The basis function of the wavelet transformation $\psi(x)$ has to satisfy the following condition

$$0 < \int_{-\infty}^{\infty} \frac{|\Psi(\omega)|^2}{|\omega|} d\omega < \infty \quad , \tag{2.40}$$

where $\Psi(\omega)$ is the Fourier transform (2.36) of the wavelet $\psi(x)$, i. e., $\Psi(\omega) = \mathrm{FT}\{\psi\}$. Using a scaling parameter $\alpha \in \mathbb{R}^+$ and a shift parameter $\beta \in \mathbb{R}$ from the basis wavelet (*mother wavelet*)

$\psi(x)$ (with $\alpha = 1$ and $\beta = 0$) a set of wavelets

$$\psi_{\alpha,\beta}(x) = \frac{1}{\sqrt{\alpha}}\psi\left(\frac{x-\beta}{\alpha}\right) \tag{2.41}$$

can be created. With the scaling parameter α the mother wavelet $\psi(x)$ can be dilated ($\alpha > 1$) or jolted ($\alpha < 1$). The shift parameter β is responsible for the translation of the mother wavelet along the argument axis. The basic idea of the wavelet transformation is to represent any arbitrary function $f(x)$ as a superposition of wavelets $\psi_{\alpha,\beta}(x)$ for different values of α and β weighted with the corresponding values of wavelet transforms $\mathrm{WT}_f(\alpha,\beta)$.

Definition 2.8 *A continuous wavelet transform* $\mathrm{WT}_f(\alpha,\beta)$ *of a on* $\mathbb{R}$ *square integrable function* $f(x) \in L^2(\mathbb{R})$ *is defined by (e. g., [Lou94, Cas96])*

$$\mathrm{WT}_f(\alpha,\beta) = \int_{-\infty}^{\infty} f(x)\psi_{\alpha,\beta}(x)dx \quad . \tag{2.42}$$

Using the inverse wavelet transformation the original function $f(x)$ can be reconstructed as follows

$$f(x) = \frac{1}{c_\psi}\int_0^{\infty}\int_{-\infty}^{\infty} \mathrm{WT}_f(\alpha,\beta)\psi_{\alpha,\beta}(x)\frac{d\beta d\alpha}{\alpha^2} \quad , \tag{2.43}$$

where c_ψ depends on the Fourier transform $\Psi(\omega)$ (2.36) of the wavelet $\psi(x)$ and can be computed with the formula

$$c_\psi = \int_0^{\infty} \frac{|\Psi(\omega)|^2}{|\omega|}d\omega = \int_{-\infty}^{0} \frac{|\Psi(\omega)|^2}{|\omega|}d\omega < \infty \quad . \tag{2.44}$$

If a function $f(x)$ has to be represented as a sum of wavelets $\psi_{\alpha,\beta}(x)$, we do not have to compute the wavelet transforms $\mathrm{WT}_f(\alpha,\beta)$ for all real values of α and β. We can limit the used scaling and shift parameters to $\alpha = 2^{-s}$ and $\beta = k\alpha$, where s and k are integer numbers ($s \in \mathbb{Z}$, $k \in \mathbb{Z}$). In this case, the set of wavelets from (2.41) can be written as

$$\psi_{s,k}(x) = \sqrt{2^s}\psi(2^s x - k) \quad . \tag{2.45}$$

In this way each square integrable function $f(x) \in L^2(\mathbb{R})$ can be expressed as a *wavelet series*

$$f(x) = \sum_{s=-\infty}^{\infty}\sum_{k=-\infty}^{\infty} d_{s,k}\psi_{s,k}(x) \quad . \tag{2.46}$$

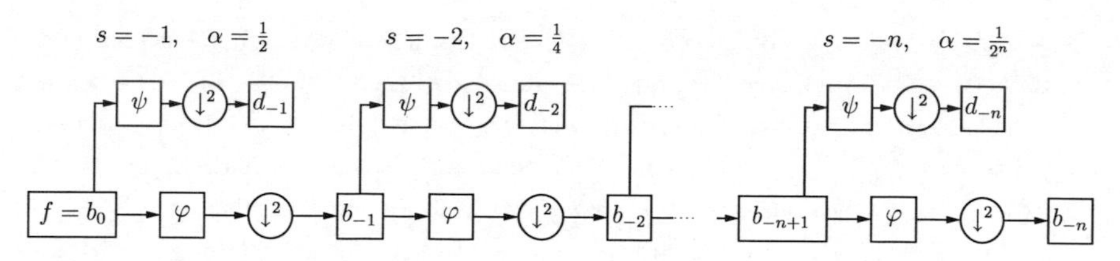

Figure 2.6: Decomposition of the one-dimensional signal f in n steps. Each signal $b_{i=0,-1,\ldots,-n+1}$ is filtered with the filters ψ and φ and downsampled by factor two ($\downarrow^2$), which divides the signal into the high-pass part d_{i-1} and low-pass part b_{i-1}.

The coefficients $d_{s,k}$ are results of the wavelet transformation $\mathrm{WT}_f(2^{-s}, k2^{-s})$ (2.42) computed as follows

$$d_{s,k} = \int_{-\infty}^{\infty} f(x)\psi_{s,k}(x)dx \quad . \tag{2.47}$$

Using the wavelets $\psi_{s,k}(x)$ and corresponding *scaling functions* $\varphi_{s,k}(x)$ the signal decomposition with the so-called *wavelet multiresolution analysis* can be introduced [Mal89]. The computation of a basis scaling function $\varphi(x)$ for a given basis wavelet $\psi(x)$ is explained later. Analogical to wavelets (2.45), the scaling functions are computed for different values of s and k

$$\varphi_{s,k}(x) = \sqrt{2^s}\varphi(2^s x - k) \quad . \tag{2.48}$$

The wavelets $\psi_{s,k}(x)$ represent high-pass filters, and the scaling functions $\varphi_{s,k}(x)$ low-pass filters in the meaning of signal processing. For example, the basis *Haar wavelet* is defined by

$$\psi(x) = \begin{cases} 1, & \text{if } \ 0 \le x < \frac{1}{2} \\ -1, & \text{if } \ \frac{1}{2} \le x < 1 \\ 0, & \text{otherwise} \end{cases} \quad , \tag{2.49}$$

and the corresponding basis scaling function by

$$\varphi(x) = \begin{cases} 1, & \text{if } \ 0 \le x < 1 \\ 0, & \text{otherwise} \end{cases} \quad . \tag{2.50}$$

The decomposition of one-dimensional signal is presented in Figure 2.6. First, the original function $f(x) = b_0$ is filtered with the wavelet ψ (high-pass filter), and the resolution of the result function reduced by factor two. The function d_{-1} contains the difference information from the original signal. The original signal $f(x) = b_0$ is also filtered with the scaling function φ (low-pass filter), and the resolution of the result function also reduced by factor two. In the function b_{-1}, the low-pass information of the original function is stored. This process is repeated for the low-pass parts $b_{i=0,-1,\ldots,-n+1}$ of the signal until the needed resolution is obtained. Each scaling function $\varphi_{s,k}(x)$ in the scale s can be expressed as a sum of scaling functions $\varphi_{s+1,k}(x)$ in the scale $s+1$. For example, a basis scaling function $\varphi(x)$ (with $s = 0$ and $k = 0$) can be written as

$$\varphi(x) = \sqrt{2^0}\varphi(2^0 x - 0) = \sqrt{2^1}\sum_k h_k \varphi(2^1 x - k) \quad , \tag{2.51}$$

where the coefficients are computed with

$$h_k = \sqrt{2}\int_{-\infty}^{\infty} \varphi(x)\varphi(2^1 x - k)dx \quad . \tag{2.52}$$

Also the wavelets $\psi_{s,k}(x)$ can be written as a weighted sum with the coefficients g_k of the scaling functions $\varphi_{s+1,k}(x)$. The basis wavelet $\psi(x)$ is equal to

$$\psi(x) = \sqrt{2^0}\psi(2^0 x - 0) = \sqrt{2^1}\sum_k g_k \varphi(2^1 x - k) \quad , \tag{2.53}$$

where

$$g_k = \sqrt{2}\int_{-\infty}^{\infty} \psi(x)\varphi(2^1 x - k)dx \quad . \tag{2.54}$$

The correspondence between the coefficients g_k of the basis wavelet $\psi(x)$ and the coefficients h_k of the basis scaling function $\varphi(x)$ is following

$$h_k = (-1)^k g_{1-k} \quad . \tag{2.55}$$

Using the above considerations, the basis wavelet $\psi(x)$ can be determined from the given basis scaling function $\varphi(x)$ and vice versa. First, with (2.52), the coefficients h_k are computed using the known function $\varphi(x)$. The wavelet function coefficients g_k result from the scaling function coefficients h_k according to (2.55). Thus, all information needed for the calculation of the basis wavelet $\psi(x)$ with the equation (2.53) are given.

2.2.3 Discrete Wavelet Transformation

In real applications, signals do not exist in continuous, but in discrete form (e. g., images). Therefore, the *discrete wavelet transformation* is very important. The scaling function φ with the coefficients h_k as well as the wavelet ψ with the coefficients g_k can be computed for different values of s and k also in the discrete case. The low-pass coefficients $b_{s,k}$ for a one-dimensional signal (see Figure 2.6) can be computed from the coefficients $b_{s+1,k}$ as follows [Chu92]

$$b_{s,k} = \sum_{i=-\left\lfloor \frac{N_\psi - 1}{2} \right\rfloor}^{\left\lceil \frac{N_\psi - 1}{2} \right\rceil} h_i b_{s+1,2k+i} \quad , \tag{2.56}$$

and the high-pass coefficients $d_{s,k}$ with

$$d_{s,k} = \sum_{i=-\left\lfloor \frac{N_\psi - 1}{2} \right\rfloor}^{\left\lceil \frac{N_\psi - 1}{2} \right\rceil} g_i b_{s+1,2k+i} \quad . \tag{2.57}$$

N_ψ is the number of the wavelet coefficients g_k and the scaling function coefficients h_k.

The decomposition of a two-dimensional signal f can be seen in Figure 2.7. First, f is filtered row by row with the scaling function φ (low-pass filter) and the wavelet ψ (high-pass filter), and the resolution of the result reduced by factor two. The low-pass coefficients come into the left side, and the high pass coefficients into the right side of the quadrant top right. Then the filtering and resolution reduction is made again column by column. The resulting quadrants have the following meaning:

- b_s: low-pass coefficients in the horizontal and vertical direction,
- $d_{0,s}$: low-pass coefficients in the horizontal and high-pass coefficients in the vertical direction,
- $d_{1,s}$: high-pass coefficients in the horizontal and vertical direction,
- $d_{2,s}$: high-pass coefficients in the horizontal and low-pass coefficients in the vertical direction.

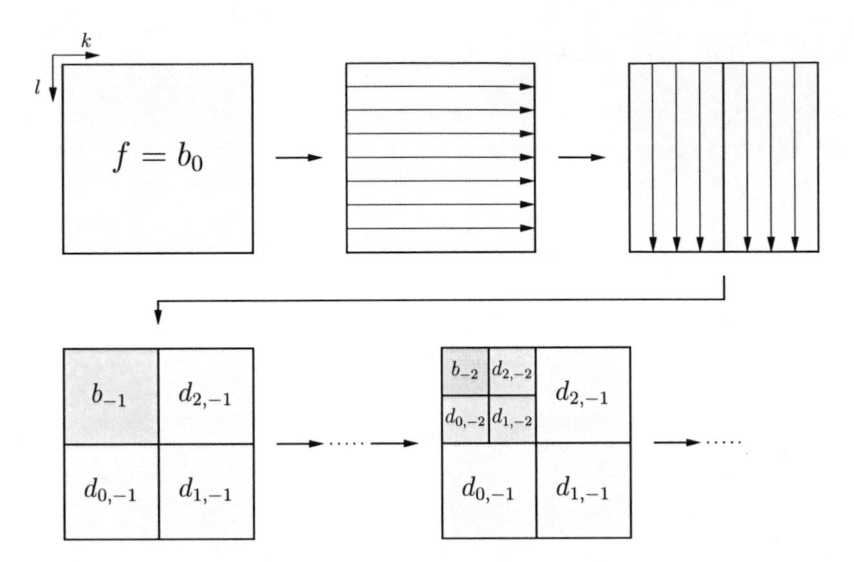

Figure 2.7: Decomposition of the two-dimensional signal f. First, f is filtered row by row with the scaling function φ (low-pass filter) and the wavelet ψ (high-pass filter), and the resolution of the result reduced by factor two. The low-pass coefficients come into the left side, and the high pass coefficients into the right side of the quadrant top right. Then the filtering and resolution reduction is made again column by column. Thus, b_s contains only low-pass coefficients from b_{s+1}, $d_{1,s}$ only high-pass coefficients, and $d_{0,s}$ and $d_{2,s}$ result from a combination of low-pass and high-pass filtering.

The coefficients in the quadrants are computed with the following formulas [Chu92]

$$b_{s,k,l} = \sum_{j=-\left\lfloor \frac{N_\psi - 1}{2} \right\rfloor}^{\left\lceil \frac{N_\psi - 1}{2} \right\rceil} h_j \sum_{i=-\left\lfloor \frac{N_\psi - 1}{2} \right\rfloor}^{\left\lceil \frac{N_\psi - 1}{2} \right\rceil} h_i b_{s+1,2k+i,2l+j} \quad , \tag{2.58}$$

$$d_{0,s,k,l} = \sum_{j=-\left\lfloor \frac{N_\psi - 1}{2} \right\rfloor}^{\left\lceil \frac{N_\psi - 1}{2} \right\rceil} g_j \sum_{i=-\left\lfloor \frac{N_\psi - 1}{2} \right\rfloor}^{\left\lceil \frac{N_\psi - 1}{2} \right\rceil} h_i b_{s+1,2k+i,2l+j} \quad , \tag{2.59}$$

$$d_{1,s,k,l} = \sum_{j=-\left\lfloor \frac{N_\psi - 1}{2} \right\rfloor}^{\left\lceil \frac{N_\psi - 1}{2} \right\rceil} g_j \sum_{i=-\left\lfloor \frac{N_\psi - 1}{2} \right\rfloor}^{\left\lceil \frac{N_\psi - 1}{2} \right\rceil} g_i b_{s+1,2k+i,2l+j} \quad , \tag{2.60}$$

$$d_{2,s,k,l} = \sum_{j=-\left\lfloor \frac{N_\psi - 1}{2} \right\rfloor}^{\left\lceil \frac{N_\psi - 1}{2} \right\rceil} h_j \sum_{i=-\left\lfloor \frac{N_\psi - 1}{2} \right\rfloor}^{\left\lceil \frac{N_\psi - 1}{2} \right\rceil} g_i b_{s+1,2k+i,2l+j} \quad . \tag{2.61}$$

2.3 Function Approximation

This section describes the main aspects of the *function approximation theory*. Detailed presentations of this topic can be found in [Che66, Dav75, Pow81, Che00]. Generally, each one-dimensional function[4] $f(x)$ defined on the set of real numbers $\mathbb{R}$ can be approximated by another function $g(x)$ defined also on $\mathbb{R}$. Using the following norm

$$||f||_2 = \sqrt{\int_{-\infty}^{\infty} |f(x)|^2 dx} \quad , \tag{2.62}$$

the so-called *best approximation* $\hat{g}(x)$ of the function $f(x)$ can be defined.

Definition 2.9 *A function $\hat{g}(x)$ is denoted as the best approximation of $f(x)$ if for all $g(x)$ defined on $\mathbb{R}$*

$$||f - \hat{g}||_2 \leq ||f - g||_2 \quad . \tag{2.63}$$

The approximation function $g(x)$ can be expressed as a weighted sum of N_υ basis functions $\upsilon_i(x)$ with the weights a_i

$$g(x) = \sum_{i=0}^{N_\upsilon - 1} a_i \upsilon_i(x) \quad . \tag{2.64}$$

The *approximation error* ε is given by

$$||f - g||_2 = ||f - \sum_{i=0}^{N_\upsilon - 1} a_i \upsilon_i||_2 = \varepsilon \quad . \tag{2.65}$$

[4]This consideration is easily extendable to N_x-dimensional functions $f(\boldsymbol{x})$ defined on $\mathbb{R}^{N_x}$ by replacement of the scalar argument x by a vector $\boldsymbol{x} = (x_1, x_2, \ldots, x_{N_x})^{\mathrm{T}}$.

The aim of function approximation is to obtain as small as possible error ε for as small as possible number of basis functions N_υ. To solve this problem the following kinds of basis functions are used in mathematics amongst others:

- polynomials,
- trigonometric functions,
- radial functions.

Polynomials are mostly applied to approximation of one-dimensional functions defined on a finite interval $x \in [x_0, x_1]$. The most elementary form of one-dimensional polynomial is called *Taylor expansion*

$$g(x) = \sum_{i=0}^{N_\upsilon - 1} a_i x^i \quad . \tag{2.66}$$

In the function approximation theory, *Legendre polynomials* and *Chebyshev polynomials* are often used. The Legendre polynomials $L_i(x)$ are defined by

$$L_i(x) = \frac{1}{i!2}\frac{d^i}{dx^i}(x^2-1)^i \quad \text{with} \quad L_0 \equiv 0 \quad , \tag{2.67}$$

and the Chebyshev polynomials by

$$T_i(x) = \cos(i \arccos x) \quad . \tag{2.68}$$

The Chebyshev polynomials can be also recursive computed with

$$T_{i+1}(x) = 2xT_i(x) - T_{i-1}(x) \quad \text{with} \quad T_0(x) = 1 \quad \text{and} \quad T_1(x) = x \quad . \tag{2.69}$$

The approximation of multi-dimensional functions $f(x_1, x_2, \ldots, x_{N_x})$ with polynomials as basis functions is seldom done.

Trigonometric functions are mostly applied to approximation of one- or two-dimensional functions. The *sine-cosine transformation* (Fourier transformation) was already introduced in Section 2.2.1. From (2.29), it follows that the approximation function $g(x)$ of the one-dimensional 2π-periodic function $f(x)$ can be expressed with (2.64) using the basis functions

$$\upsilon_i(x) = \begin{cases} 1, & \text{if} \quad i = 0 \\ \cos\left(\frac{i+1}{2}x\right), & \text{if} \quad i = 2j-1 \\ \sin\left(\frac{i}{2}x\right), & \text{if} \quad i = 2j \end{cases} \quad , \tag{2.70}$$

where $i, j \in \mathbb{N}$. If the period length X_P of the function is not equal to 2π, then the argument x in (2.70) has to be replaced by $\frac{2\pi}{X_P}x$. It is also possible to use only cosine terms to approximate functions. The *cosine transformation* modifies (2.64) to

$$g(x) = \sum_{i=0}^{N_\upsilon - 1} a_i \cos\left(\frac{i\pi}{X_P}x\right) \quad . \tag{2.71}$$

Both sine-cosine and cosine transformation are easily extendable to two dimensions. In the two-dimensional case, the cosine transformation is expressed with

$$g(x, y) = \sum_{i=0}^{N_{\upsilon y}-1} \sum_{j=0}^{N_{\upsilon x}-1} a_{iN_{\upsilon x}+j} \cos\left(\frac{j\pi}{X_P}x\right) \cos\left(\frac{i\pi}{Y_P}y\right) \quad , \tag{2.72}$$

where $N_{\upsilon x}$ and $N_{\upsilon y}$ are numbers of basis functions in x- and y-direction respectively, and Y_P is the period length in the y-direction.

Apart from polynomials and trigonometric functions, the function approximation can be performed also with radial functions. A fundamental property of the radial basis functions $\upsilon_i(x)$ is that they have their own central points μ_i, i. e., the function values depend on the argument distance from the central point $|x - \mu_i|$. Therefore, the basis functions are denoted by $\upsilon_i(|x - \mu_i|)$. In this case, the approximation function (2.64) is given with the following formula

$$g(x) = \sum_{N_\upsilon - 1}^{i=0} a_i \upsilon_i(|x - \mu_i|) \quad , \tag{2.73}$$

where the basis functions $\upsilon_i(|x - \mu_i|)$ can be, e. g., equal to

$$\upsilon_i(|x - \mu_i|) = \begin{cases} |x - \mu_i| \\ |x - \mu_i|^3 \\ \exp(-|x - \mu_i|^2) \\ (|x - \mu_i|^2 + b^2)^{-\frac{1}{2}} \\ (|x - \mu_i|^2 + b^2)^{\frac{1}{2}} \end{cases} \quad . \tag{2.74}$$

For the function approximation with radial functions, not only the weights a_i, but also the central points μ_i have to be determined.

In real applications, the values of the function $f(x)$, which has to be approximated, are only known for the finite number N_T of points $(f(x_0), f(x_1), \ldots, f(x_{N_T-1}))$. Using this knowledge and (2.64), the following system of equations can by constructed

$$\begin{cases} f(x_0) - \sum\limits_{i=0}^{N_\upsilon-1} a_i \upsilon_i(x_0) = 0 \\ f(x_1) - \sum\limits_{i=0}^{N_\upsilon-1} a_i \upsilon_i(x_1) = 0 \\ \quad \dots \\ f(x_{N_T-1}) - \sum\limits_{i=0}^{N_\upsilon-1} a_i \upsilon_i(x_{N_T-1}) = 0 \end{cases} . \tag{2.75}$$

If $N_T \geq N_\upsilon$ and the equations are linearly independent, then the weights $a_{i=0,1,\dots,N_\upsilon-1}$ can be determined with the equation system above (2.75). In this way, the function values $f(x)$ can be precisely interpolated for all $x \in [x_0, x_{N_T}]$, and sometimes even for $x \in \mathbb{R}$.

2.4 Summary

In this chapter, mathematical instruments were discussed, which are used for the development of the object recognition system.

Section 2.1 dealt with fundamentals of mathematical statistics. First, basic terms like random variable (Definition 2.1), its distribution (Definition 2.2), probability density function (Definition 2.3), mean value (Definition 2.4), variance (Definition 2.5), and covariance matrix (2.16) were defined. Then two specific kinds of random variables, normal and uniform, were introduced with their density functions (2.19), (2.20), and (2.22). After that, the join density function for two random variables (2.24) was taken into consideration. At the end, the basic idea of the maximum a-posteriori estimation (MAP) (2.28) was presented.

In Section 2.2, the theory of wavelet transformation was introduced. First (Section 2.2.1), the main differences between Fourier signal analysis and the wavelet signal processing were described. Then in (Section 2.2.2), the wavelet transformation for continuous functions, and the multiresolution analysis of one-dimensional signals were shown. The last part (Section 2.2.3) extended the considerations to two-dimensional discrete case, which is very important for the present work, because images are two-dimensional discrete signals.

Finally, in Section 2.3, the basic idea of the function approximation theory was shortly explained. After defining the best approximation function $g(x)$ for an arbitrary function $f(x)$ (Definition 2.9), three different kinds of approximation basis functions, namely polynomials, trigonometric functions, and radial functions, were presented. At the end of this section was shown how to interpolate a function knowing only the finite number of its values (2.75).

Chapter 3

Algorithms and Methods for Object Recognition

This chapter presents known algorithms and methods for object classification and localization. According to the general scheme of a system for object recognition depicted in Figure 1.3, Section 3.1 starts with the *image acquisition* process. Then it describes main properties of image databases for evaluation of object recognition systems and discusses some algorithms for image preprocessing. Finally, it presents the general tasks of object feature extraction, object modeling, and object recognition independent on the recognition system type. Concerning feature extraction, the approaches for object recognition can be divided into shape-based and appearance-based. The shape-based algorithms with all their steps, shape representation, modeling, and object classification and localization are described in Section 3.2. Section 3.3 starts with the feature extraction in the case of appearance-based methods. After that, it considers some known terms in appearance-based object recognition like: *template matching*, *eigenspace approach*, *neural networks*, and *support vector machines*. Section 3.4 closes the present chapter with a summary.

3.1 Fundamentals

In order to classify and localize objects in digital images, the image data should be first acquired and preprocessed (see Figure 1.3). This section starts with these beginning steps of the object recognition process (Section 3.1.1 and Section 3.1.3) as well as some general properties of image databases (Section 3.1.2). Moreover, it presents the general tasks of object feature extraction (Section 3.1.4), object modeling, and recognition (Section 3.1.5).

3.1.1 Image Acquisition

The image acquisition process is made by a camera. A camera can be mathematically denoted as a mapping of points from a three-dimensional world coordinate system into the two-dimensional image. Each point in the world coordinate system can be expressed with coordinates in the camera coordinate system. There are several camera projection models, in which a 3D point can be mapped from the camera coordinate system into the coordinate system of the image. Some commonly used methods are described in the following.

Perspective Projection

A 3D point from the camera coordinate system can be projected onto the image plane by a *perspective projection* (e. g., [Pau03, Ma05]). Using homogeneous coordinates, this mapping can be written as a linear projection matrix. The entries of this matrix are called *intrinsic* camera parameters, and represent the physical properties of the camera. The perspective projection is a simplification of the true behavior of a real camera, but it is sufficient for most object recognition systems. The geometric model for perspective projection is the so-called *pinhole camera* (see Figure 3.1). Given $\boldsymbol{p} = (p_x, p_y, p_z)^{\mathrm{T}}$ and focal length f the coordinates of the image point $\boldsymbol{p}' = ({p_x}', {p_y}')^{\mathrm{T}}$ can be computed using the Tales' theorem. It means that

$$\frac{{p_x}'}{p_x} = \frac{f}{p_z} \quad \Leftrightarrow \quad {p_x}' = \frac{f}{p_z} p_x \quad , \tag{3.1}$$

and

$$\frac{{p_y}'}{p_y} = \frac{f}{p_z} \quad \Leftrightarrow \quad {p_y}' = \frac{f}{p_z} p_y \quad . \tag{3.2}$$

The farther $\boldsymbol{p}$ is away from the image plane, the closer its projection $\boldsymbol{p}'$ will be to the image center. On the other hand, the larger the focal length f is, the farther the projection will be from the principal point. These non-linear equations (3.1) and (3.2) become linear using homogeneous coordinates

$$\underline{\boldsymbol{p}}' = \begin{pmatrix} f & 0 & 0 & 0 \\ 0 & f & 0 & 0 \\ 0 & 0 & 1 & 0 \end{pmatrix} \underline{\boldsymbol{p}} \quad , \tag{3.3}$$

where $\underline{\boldsymbol{p}}$ and $\underline{\boldsymbol{p}}'$ are points $\boldsymbol{p}$ and $\boldsymbol{p}'$ (see Figure 3.1) expressed in homogeneous coordinates.

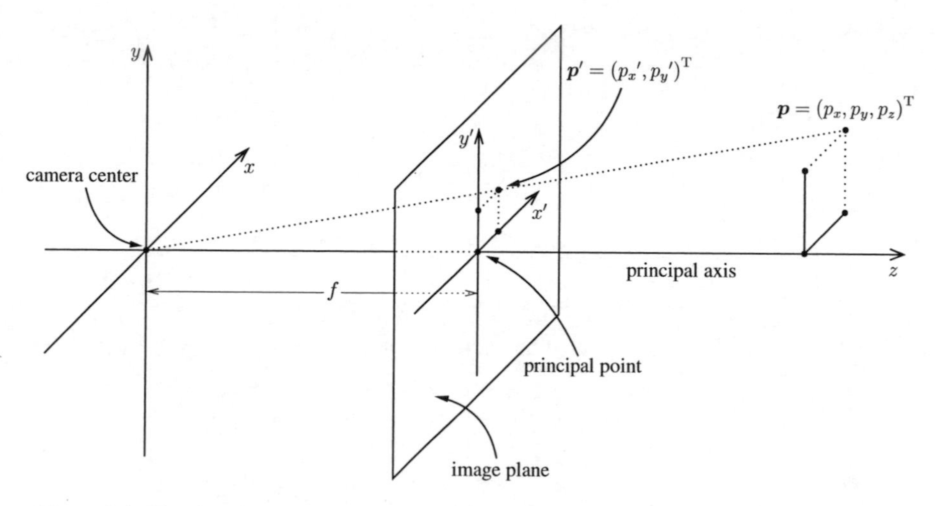

Figure 3.1: The pinhole camera model. The 3D point $\boldsymbol{p}$ in camera coordinates is projected onto the image plane, which yields $\boldsymbol{p}'$ in image coordinates. Given $\boldsymbol{p} = (p_x, p_y, p_z)^{\mathrm{T}}$ and focal length f, the coordinates of the image point $\boldsymbol{p}' = (p_x', p_y')^{\mathrm{T}}$ can be computed using the Tales' theorem.

Orthographic Projection

The most simple camera projection model which is used for image acquisition, is the so called *orthographic projection* (e. g., [Pau03]). It simplifies equations (3.1) and (3.2) into

$$\boldsymbol{p}' = \boldsymbol{p}'_1 = \begin{pmatrix} p_x \\ p_y \end{pmatrix} \quad , \tag{3.4}$$

which is illustrated in Figure 3.2. The z-coordinate of the 3D point is ignored and the point is projected parallel to the principal axis onto the image plane. Orthographic projection is a special case of perspective projection, with $f \to \infty$ and $p_z \to \infty$. The perspective projection equation for homogeneous coordinates (3.3) changes in the case of orthographic projection to

$$\underline{\boldsymbol{p}}' = \begin{pmatrix} 1 & 0 & 0 & 0 \\ 0 & 1 & 0 & 0 \\ 0 & 0 & 0 & 1 \end{pmatrix} \underline{\boldsymbol{p}} \quad . \tag{3.5}$$

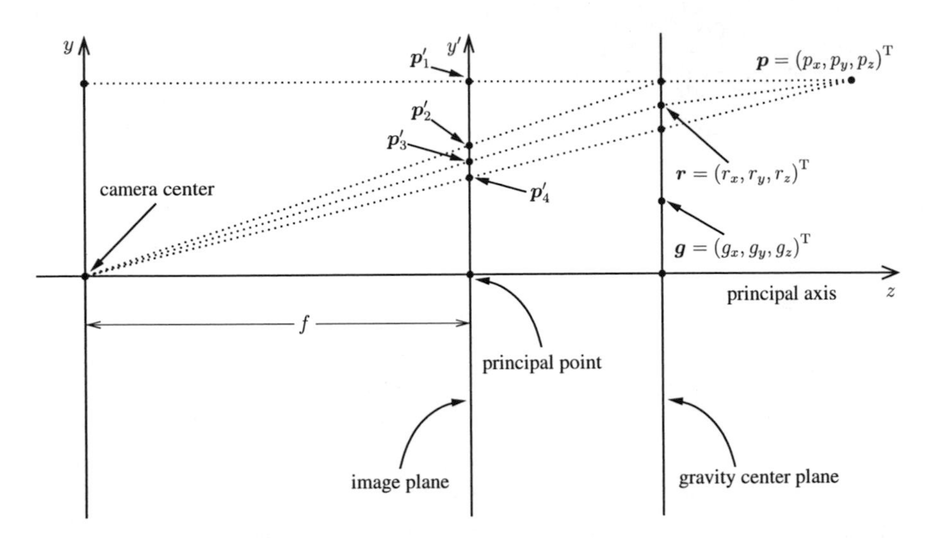

Figure 3.2: Projection of a 3D point $\boldsymbol{p}$ with different projection models. $\boldsymbol{p}'_1$ is the orthographic projection, $\boldsymbol{p}'_2$ the weak perspective projection, $\boldsymbol{p}'_3$ the paraperspective projection, and $\boldsymbol{p}'_4$ the perspective projection of $\boldsymbol{p}$ onto the image plane. $\boldsymbol{g}$ denotes the gravity center of all 3D points.

Weak Perspective Projection

Weak perspective projection is a combination of the orthographic and the perspective projection model (e. g., [Tru98]). First, a 3D point $\boldsymbol{p}$ given in the camera coordinates is orthographically projected onto a plane, which is parallel to the image plane and contains the gravity center $\boldsymbol{g}$ of all 3D points (see Figure 3.2). Then it follows the perspective projection and gives the result point $\boldsymbol{p}'_2$ on the image plane. For the weak perspective projection equations (3.1) and (3.2) change into

$$\boldsymbol{p}' = \boldsymbol{p}'_2 = \frac{f}{g_z} \begin{pmatrix} p_x \\ p_y \end{pmatrix} \quad . \tag{3.6}$$

Using homogeneous coordinates, the formula (3.6) can be written as

$$\underline{\boldsymbol{p}}' = \begin{pmatrix} f & 0 & 0 & 0 \\ 0 & f & 0 & 0 \\ 0 & 0 & 0 & g_z \end{pmatrix} \underline{\boldsymbol{p}} \quad . \tag{3.7}$$

In equation (3.6), the z-coordinate g_z of the gravity center $\boldsymbol{g}$ instead of the true point depth p_z (as in (3.1) and (3.2)) is used. It means that each 3D point $\boldsymbol{p}$ is treated as having the same depth in the case of weak perspective projection. The weak perspective projection model as an approximation of the real projection is only valid under certain conditions. The maximal depth difference between the points in the scene has to be at least twenty times smaller than their average distance from the camera.

Paraperspective Projection

The last projection model which is introduced in this section is the *paraperspective projection* (e. g., [Alo90]). The difference to the weak perspective projection is that in the first step, the 3D point $\boldsymbol{p}$ is not orthographically projected onto the gravity center plane, but projected parallel to the line connecting the gravity center $\boldsymbol{g}$ with the camera center (see Figure 3.2). The projection of the 3D point $\boldsymbol{p}$ onto the gravity center plane is given by

$$\boldsymbol{r} = \boldsymbol{p} - \frac{p_z - g_z}{g_z}\boldsymbol{g} = \boldsymbol{p} - \alpha\boldsymbol{g} \quad . \tag{3.8}$$

In the second step of the paraperspective projection, $\boldsymbol{r}$ is perspective projected onto the image plane, which yields $\boldsymbol{p}'_3$ (see Figure 3.2). This can be expressed according to (3.1) and (3.2) as follows

$$\boldsymbol{p}' = \boldsymbol{p}'_3 = \frac{f}{r_z}\begin{pmatrix} r_x \\ r_y \end{pmatrix} \quad . \tag{3.9}$$

Equations (3.8) and (3.9) result in the paraperspective projection equation

$$\boldsymbol{p}' = \boldsymbol{p}'_3 = \frac{f}{g_z}\begin{pmatrix} p_x - \alpha g_x \\ p_y - \alpha g_y \end{pmatrix} \quad . \tag{3.10}$$

The paraperspective projection can also be written using homogeneous coordinates

$$\underline{\boldsymbol{p}}' = \begin{pmatrix} f & 0 & 0 & -f\alpha g_x \\ 0 & f & 0 & -f\alpha g_y \\ 0 & 0 & 0 & g_z \end{pmatrix} \underline{\boldsymbol{p}} \quad . \tag{3.11}$$

The advantage of the paraperspective projection in comparison to the perspective projection is that objects with large depth differences can also be modeled.

3.1.2 Image Databases

With the methods described in Section 3.1.1, image databases are acquired which are needed for the training and testing of object recognition systems. The first information about an image database is the set of objects (object classes)

$$\Omega = \{\Omega_1, \Omega_2, \ldots, \Omega_\kappa, \ldots, \Omega_{N_\Omega}\} \quad , \tag{3.12}$$

which are taken into account for the classification task. The complexity of the classification problem defined in Section 1.1 depends on the choice (object similarity) and the number of objects in the database. The degree of difficulty of the localization task (Section 1.1) depends on the number and type of the transformation parameters (internal or external) which define the object pose (see Figure 1.2) in the image database. The object pose can be described with up to six transformation parameters

$$\boldsymbol{t} = (t_x, t_y, t_z)^{\mathrm{T}} \quad \text{and} \quad \boldsymbol{\phi} = (\phi_x, \phi_y, \phi_z)^{\mathrm{T}} \quad . \tag{3.13}$$

Under 2D image database, a set of object images is understood, where the objects are rotated and translated with the internal pose parameters ($\phi_{\text{int}} = \phi_z$, $\boldsymbol{t}_{\text{int}} = (t_x, t_y)^{\mathrm{T}}$). In a 3D image database, objects are transformed with the external pose parameters ($\boldsymbol{\phi}_{\text{ext}} = (\phi_x, \phi_y)^{\mathrm{T}}$, $t_{\text{ext}} = t_z$), which can change their size and appearance (see Figure 1.2).

The most popular image database for object recognition is the COIL (Columbia Object Image Library) database with gray level images. COIL-20 presented in [Nen96b] consists of 20 objects which are depicted in Figure 3.3. COIL-100 [Nen96a] is a completion of COIL-20 with additional 80 objects. However, the COIL image database has some disadvantages, which make an objective evaluation of an object recognition system impossible:

- *Object Pose*
  COIL has only one degree of freedom, i. e., the object pose in COIL is defined only with one external transformation parameter, namely the external rotation ϕ_y (see Figure 1.2). Thus, the degree of difficulty of the localization problem is low for this database.

- *Illumination*
  All images in COIL were taken under identical illumination conditions, which significantly simplifies the object recognition task. Such a database cannot be applied to evaluation of systems, which have to classify and localize objects in real world environments, where the illumination varies.

Figure 3.3: Example images of all 20 objects of the Columbia Object Image Library COIL-20 [Nen96b]

- *Size Normalization*
  An object size normalization was introduced in COIL, i. e., the problem with the variable object area size, which occurs in real applications of object recognition systems, is not taken into account.

- *Brightness Normalization*
  A brightness normalization was also introduced in COIL. This means that the value of the brightest pixel in each image was set to 255, and the values of remaining pixels were scaled with the same factor. It improves the image contrast and simplifies the recognition problem in comparison to real situations.

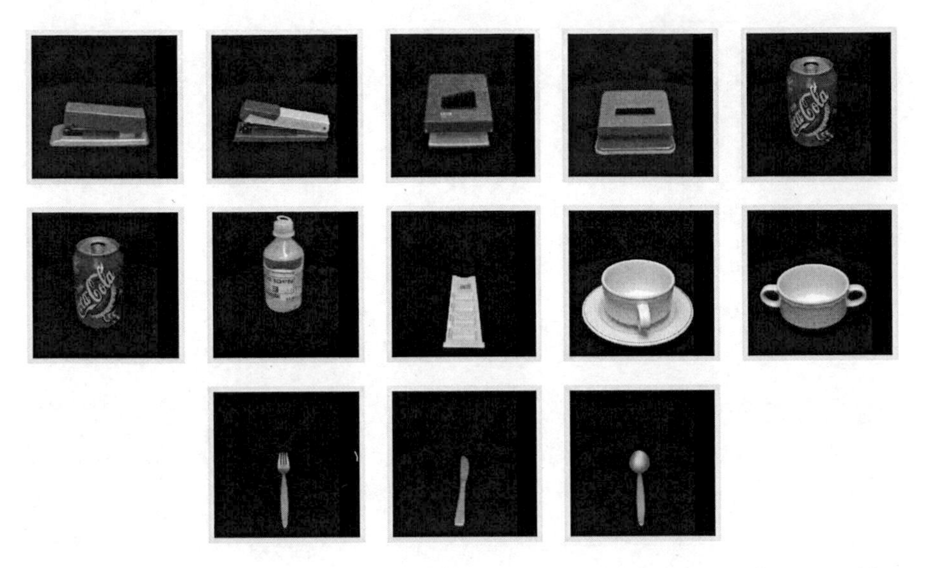

Figure 3.4: Example images of all 13 objects of the image database for 3D object recognition DIROKOL [Rei01].

- *Homogeneous Background*
  In all images, objects are placed on homogeneous (black) background. There are no real test images with heterogeneous background and occlusions in COIL. Thus, an objective evaluation of an object recognition system, which has to work in a real environment, seems to be impossible using COIL.

Another image database which can be applied to evaluation of systems for computational classification and localization of objects in digital images, is the image database for 3D object recognition DIROKOL [Rei01]. This database consists of 13 objects from the office and health care domain (see Figure 3.4). DIROKOL allows more objective evaluation of object recognition systems than COIL, but it makes still some assumptions about the experimental environment. The main properties of DIROKOL are following:

- *Object Pose*
  DIROKOL has four degrees of freedom. The object pose in DIROKOL is defined with external rotations $\boldsymbol{\phi}_{\text{ext}} = (\phi_x, \phi_y)^{\mathrm{T}}$ and internal translations $\boldsymbol{t}_{\text{int}} = (t_x, t_y)^{\mathrm{T}}$ (see Figure 1.2). The localization task seems to be more complex as in the case of COIL image database.

- *Illumination*

  The images were taken under three different illumination conditions, which gives the possibility to evaluate an object recognition system for variable illumination.

- *Artificial Test Images*

  Unfortunately, DIROKOL does not comprehend test images with real heterogeneous background and occlusions. In [Rei04], artificially created test images are used. First, the objects were taken separate from the images with background, and then, they were artificially pasted on the background and sometimes occluded with a special image processing tool. The creation of the test images in this way does not take into account that the object appearance (object pixel values) depends also on its neighborhood due to real occlusions, illumination reflections, or shadows. Using such test images is a simplification of the object recognition problem.

Due to the listed restrictions given by the existing image databases for object recognition, a new image database for 3D object recognition in real world environment 3D-REAL-ENV was generated within the scope of the present work. The detailed description of this database follows in Section 6.2.

3.1.3 Image Preprocessing

Generally, the image acquisition process (Section 3.1.1) yields true color images (RGB images). However, before the images are used for object feature extraction, they are first preprocessed (see Figure 1.3). There are several possibilities to preprocess true color images. Usually the original images are converted into the gray level images (e. g., [Rei04]) and resized (e. g., [Rei05]). Sometimes object feature vectors are computed based on image segmentation results like lines or corners (e. g., [Ker03, Ces05]), which is characteristic for shape-based approaches. In this case the original images can be preprocessed with different pixel based or edge based segmentation algorithms. A detailed consideration of *digital image processing* which can be found e. g., in [Cas96, Gon92, Pra01], is not the objective of this work; thus, only the basic idea of the preprocessing algorithms is given in this section.

From True Color to Gray Level Images

True color images (RGB images) are images in which each pixel is specified by three values, namely the red, green, and blue components of the pixel color. Graphics file formats store true color images usually as 24 bit images, where the red, green, and blue channels are defined with 8 bits. This yields a potential of more than 16 million colors. In contrast to this, gray level images (intensity images) represent each pixel with only one value, the gray level intensity, which is mostly stored using 8 bits. Generally, each mathematical operation, which transforms a $n \times m \times 3$ sized matrix into a $n \times m \times 1$ sized matrix, can be considered as a conversion of a RGB image of size $n \times m$ pixels into a gray level image of the same size. However, in practice, this conversion is made locally (pixel by pixel). This means that the intensity of each pixel in a gray level image is computed using the red, green, and blue components of the corresponding pixel in a RGB image, e. g., by taking the mean value of these components.

Image Resizing

Before images are used for object recognition, they are often resized. Due to the minimization of the computing time of the object feature vectors, the original images can be scaled down. Generally, each operation which converts an image of size $n \times m$ pixels into an image of size $n' \times m'$ pixels, where $n \neq n'$ or $m \neq m'$, can be denoted as an image resizing algorithm. The fundamental question of an image resizing algorithm is: How to get the $n' \times m'$ pixels from the $n \times m$ pixels? The simplest idea of the image scaling down algorithms ($n' < n$, $m' < m$) is the omission of $nm - n'm'$ pixels from the original image, which are uniformly distributed over the whole original image. Other algorithms compute values of the $n'm'$ pixels using interpolation of pixel values from the original image. If the converted image has to be bigger than the original ($n' > n$, $m' > m$), then there is a problem to produce additional $n'm' - nm$ pixels without any additional information. Most algorithms just duplicate the missing $n'm' - nm$ pixels from the original image in this case.

Segmentation

The goal of *digital image segmentation* is the division or separation of an image into regions of similar attribute. The most basic attribute for segmentation is the pixel grayscale value for a gray level image or values of pixel color components for a color image. Image edges and texture are also useful attributes for segmentation. Haralick and Shapiro [Har85] have established the following qualitative guideline for a good image segmentation: "Regions of an image segmentation should be uniform and homogeneous with respect to some characteristics such as gray tone or texture. Region interiors should be simple and without many small holes. Adjacent regions of a segmentation should have significantly different values with respect to the characteristic on which they are uniform. Boundaries of each segment should be simple, not ragged, and must be spatially accurate." Unfortunately, no quantitative image segmentation performance metric has been developed. A detailed description of image segmentation methods, which can be found, e. g., in [Fu81, Har85, Pal93], is not the objective of the present work. Thus, only a short survey of existing segmentation algorithms is given in the following. Methods for image segmentation can be divided into:

- *Amplitude Segmentation Methods*
  Amplitude segmentation methods are based on the thresholding of pixel grayscale values for gray level images or the thresholding of pixel color component values in the case of color images. Many images can be characterized as containing some object of interest of reasonably uniform brightness placed against a background of different brightness. Typical examples are, e. g., handwritten text or microscope biomedical samples. For such images, luminance is a distinguishing feature that can be used to segment the object from the background. Practical problems occur, however, when the observed image is subject to noise and when both the object and background assume some broad range of pixel values. Another frequent difficulty is that the background may be not homogeneous.

- *Clustering Segmentation Methods*
  One of the earliest examples of image segmentation using data clustering was introduced by Haralick and Kelly [Har69]. They subdivided multispectral aerial images of agricultural land into regions containing the same type of land cover. The clustering segmentation concept is simple. Each pixel in the image is represented by a vector of feature measurements $\boldsymbol{c} = (c_1, c_2, \ldots, c_{N_c})^{\mathrm{T}}$. If the measurement set has to be effective for image segmentation, data collected at various pixels within a segment of common attribute should be similar, i. e., the data should be tightly clustered in an N_c-dimensional measurement space. If this condition holds, the segmenter design task becomes one of subdividing the N_c-dimensional measurement space into mutually exclusive compartments, each of which envelopes typical data clusters for each image segment. Figure 3.5 illustrates this concept for two dimensions. If a measurement vector for a pixel falls within a measurement space compartment, the pixel is assigned the segment name (label) of that compartment.

- *Region Segmentation Methods*
  The amplitude and clustering methods are based on point properties of an image. The logical extension, as first suggested by Muerle and Allen [Mue68], is the utilization of spatial properties of an image for segmentation, e. g., using *region growing, split and merge,* and *watershed* segmentation techniques. Region growing is one of the conceptually simplest

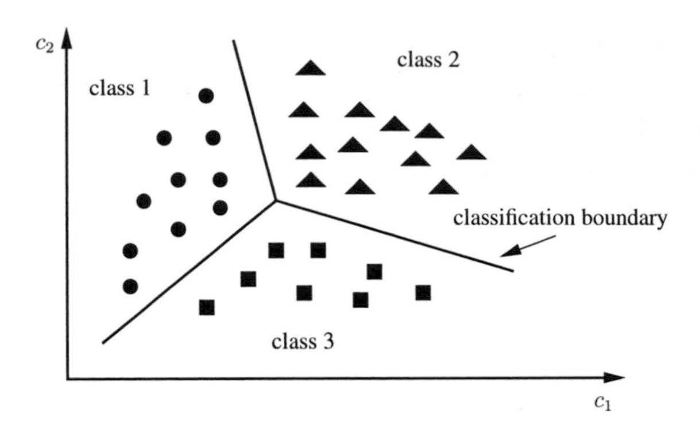

Figure 3.5: Data clustering for two feature measurements. The classification boundary splits the features into three different classes.

approaches to image segmentation, in which neighboring pixels of similar amplitude are grouped together to form a segment region [Bri70]. Split and merge image segmentation techniques are based on a quad tree data representation, whereby a square image segment is broken (split) into four quadrants if the original image segment is nonuniform in attribute. If four neighboring squares are found to be uniform, they are replaced (merged) by a single square composed of the four adjacent squares [Pav82]. Watershed segmentation techniques consider an image to be an altitude surface, in which high amplitude pixels correspond to ridge points, and low amplitude pixels correspond to valley points [Har83]. If a drop of water falls on any point of the altitude surface, it moves to a lower altitude until it reached a local altitude minimum. The accumulation of water in the neighborhood of a local minimum is called a *catchment basin*. All points that drain into a common catchment basin are part of the same watershed and build one segment.

- *Boundary Detection*

  It is possible to segment an image into regions of common attribute by detecting the boundary of each region, for which a significant change in attribute across the boundary occurs. In the first step, an edge detector (e. g., Canny operator [Can86]) is applied. If an image is noisy or if its region attributes differ by only a small amount between regions, a detected boundary may often be broken. In this case, different *edge linking* techniques can be employed to bridge short gaps in such a region boundary.

- *Texture Segmentation*
  Texture is also a valuable feature for image segmentation. One approach to texture segmentation, fostered by Rosenfeld [Ros71], is the computation of some texture coarseness measure at all image pixels and the detection of its changes. Zucker [Zuc75] proposed a histogram thresholding method of texture segmentation.

The result of any successful image segmentation is the labeling of each pixel that lies within a specific distinct segment, i. e., to each pixel of an image a label number or index of its segment is appended.

3.1.4 Object Feature Extraction

As can be seen in Figure 1.3, after the images are acquired and preprocessed, they are used for object feature extraction. Features represent the characteristics of objects, and selecting or synthesizing effective composite features are the key to the performance of object recognition. It could be imagined to use directly the image pixels $\boldsymbol{f}_{x,y}$ [1] in order to describe an object placed in the image $\boldsymbol{f}$, but this way would be rather ineffective. First, the amount of data (i. e., the size of object models) would be very large. Second, such an object representation would be very sensitive to illumination changes, object occlusions, heterogeneous background, and image noise. Third, in such a feature space, the objects would not be easily distinguishable from each other. Therefore, in practical applications, an explicit computation of the object feature vectors, which essentially reduces the amount of data, is applied. Depending on the approach, an object can by represented by one global feature vector $\boldsymbol{c}$, or N_C local feature vectors $\boldsymbol{c}_m$, which build a set of feature vectors

$$C = \{\boldsymbol{c}_1, \boldsymbol{c}_2, \ldots, \boldsymbol{c}_m, \ldots, \boldsymbol{c}_{N_C}\} \quad . \tag{3.14}$$

Each of these feature vectors can comprise one or more components. In the case of a global feature vector, it is denoted as

$$\boldsymbol{c} = (c_1, c_2, \ldots, c_q, \ldots, c_{N_c})^{\mathrm{T}} \quad . \tag{3.15}$$

Each of the local feature vectors are expressed accordingly as

$$\boldsymbol{c}_m = (c_{m,1}, c_{m,2}, \ldots, c_{m,q}, \ldots, c_{m,N_c})^{\mathrm{T}} \quad . \tag{3.16}$$

The features have to satisfy two conflicting conditions. On the one hand, they should describe

[1] Generally, the image pixels are represented by a vector $\boldsymbol{f}_{x,y}$, but for gray level images it is a scalar $f_{x,y}$.

the object as detailed as possible to allow the system a correct classification and localization. On the other hand, the features should be robust, i.e., their values should not be sensitive to illumination changes, perspective distortion, or small noise. Furthermore, it should be taken into account that for the training phase (see Figure 1.3), only a limited number of views of an object is available. The viewpoints in the recognition phase are usually different from the training views. Therefore, the values of feature vectors for adjacent viewpoints have to be similar. These and other requirements make the selection of an appropriate method for feature extraction very difficult.

The main advantage of using many local feature vectors $\boldsymbol{c}_m$ to describe an object in comparison to a description with one global feature vector $\boldsymbol{c}$ is that local disturbances, e. g., noise or occlusion, affect only the local feature vectors in a small region around it. All other local feature vectors remain unchanged. In contrast to this, a global feature vector can totally change, if only one pixel in the image varies.

The feature extraction in the case of shape-based approaches is different from the feature computation for appearance-based recognition systems. The first object recognition methods apply a segmentation process and use geometric features like lines or vertices. An overview over these methods follows in Section 3.2.1. The second type of methods calculate features directly from the pixel values without a previous segmentation step (Section 3.3.1).

3.1.5 Object Modeling and Recognition

Before objects can be classified and localized, object models $\mathcal{M}_\kappa$ for all possible object classes Ω_κ are learned in the so-called training phase (see Figure 1.3) using the feature vectors introduced in the previous section. For this purpose, a representative image database, which contains images of the objects from different viewpoints, is acquired (Section 3.1.2). Illumination changes can also be taken into account in the training phase. Though only a limited number N_ρ of training images can be acquired in the training phase, the object model $\mathcal{M}_\kappa$ should be able to handle in the recognition phase also viewpoints, which lie between the training views. The illumination in the recognition phase can also be different from the training illumination. Due to these aspects, the object modeling problem becomes very difficult and important for the object classification and localization task. The method, which is applied for object modeling, depends, on the one hand, on the type of used features, on the other hand, on the definition of a concrete object recognition problem. Generally, the object modeling for 3D object recognition is more difficult than the learning phase in the case of 2D objects. In the 3D object recognition tasks, the object pose is also defined with external pose parameters $(\boldsymbol{\phi}_{\text{ext}}, \boldsymbol{t}_{\text{ext}})$, which change the size and appearance of the objects. If the recognition system should only classify objects, the object model $\mathcal{M}_\kappa$ has to contain only the object class Ω_κ. However, for object localization, the pose parameters $(\boldsymbol{\phi}, \boldsymbol{t})$ for each training image should also be stored in the models $\mathcal{M}_\kappa$. There exist three ways to model a 3D object with varying size and appearance:

- *Without Pose Parameters [Dah99]*
  The system is only able to classify objects. Object localization is not possible in this case.

- *Discrete Pose Parameter Domain [Sch96]*
  This method uses so-called *view classes*. Adjacent training viewpoints, for which the object appearance is similar, build one view class. The object model $\mathcal{M}_\kappa$ for each view class is trained separately, i. e., each object class Ω_κ is described by multiple view classes. Thus, the object model $\mathcal{M}_\kappa$ consists of multiple object models $\mathcal{M}_{\kappa,\upsilon}$ for all view classes. In the recognition phase, the pose of an object can be estimated only approximately, i. e., the system can only determine, to which view class the object pose belongs.

- *Continuous Pose Parameters Domain [Rei05]*
  In this case the object appearance is modeled with continuous pose parameters. The object model can be expressed as a function defined on continuous pose parameter domain

$$\mathcal{M}_\kappa = \mathcal{M}_\kappa(\boldsymbol{\phi}, \boldsymbol{t}) \quad . \tag{3.17}$$

  In this way, the object pose parameters can be exactly determined.

In all of these three cases, features can be stored in models directly using their values or statistically modeled with density functions $p(C|\mathcal{M}_\kappa)$. Nowadays, statistical approaches are preferred, because they can better handle noise, illumination changes, heterogeneous background, and occlusions in real world environment.

The object recognition task can be described as follows. An image $\boldsymbol{f}$ is taken and a global feature vector $\boldsymbol{c}$ or a set of local feature vectors $C = \{\boldsymbol{c}_1, \boldsymbol{c}_2, \ldots, \boldsymbol{c}_m, \ldots, \boldsymbol{c}_{N_C}\}$ (3.14) is computed. In the following, it is assumed that a global feature vector $\boldsymbol{c}$ builds a set of feature vectors C with one element. The objective of the object classification and localization task is to find out which of the learned objects Ω_κ and in which poses $(\boldsymbol{\phi}, \boldsymbol{t})$ correspond to the extracted set of feature vectors C from the image $\boldsymbol{f}$. If statistical modeling is not performed, there are two methods to solve the recognition problem. The first one maximizes a measure which expresses how good the match between the feature vectors from the set C and the learned object models $\mathcal{M}_\kappa$ is. The second one minimizes a measure which describes the match error between C and

$\mathcal{M}_\kappa$. Using a distance function $d(C, \mathcal{M}_\kappa)$ the classification problem without localization can be described by the following equation

$$\widehat{\kappa} = \underset{\kappa}{\operatorname{argmin}}\, d(C, \mathcal{M}_\kappa) \quad . \tag{3.18}$$

Considering both classification and localization equation 3.18 changes into

$$(\widehat{\kappa}, \widehat{\boldsymbol{\phi}}, \widehat{\boldsymbol{t}}) = \underset{(\kappa,\boldsymbol{\phi},\boldsymbol{t})}{\operatorname{argmin}}\, d(C, \mathcal{M}_\kappa(\boldsymbol{\phi}, \boldsymbol{t})) \quad . \tag{3.19}$$

Statistical algorithms for object recognition do not use any distance functions. They maximize a density function $p(\Omega_\kappa|C)$, which expresses the occurrence likelihood of the object class Ω_κ in the scene $\boldsymbol{f}$ for the given set of feature vectors C. According to the maximum a-posteriori estimation (2.28) presented in Section 2.1 and concerning only the classification task, it can be written

$$\widehat{\kappa} = \underset{\kappa}{\operatorname{argmax}}\, p(\Omega_\kappa|C) = \underset{\kappa}{\operatorname{argmax}}\, p(\Omega_\kappa) p(C|\mathcal{M}_\kappa) \quad . \tag{3.20}$$

Assuming that the probability of the object occurrence in the scene $\boldsymbol{f}$ is the same for each object class Ω_κ (uniform distribution), i. e.,

$$p(\Omega_1) = \cdots = p(\Omega_\kappa) = \cdots = p(\Omega_{N_\Omega}) \tag{3.21}$$

the classification problem (3.20) can be reduced into

$$\widehat{\kappa} = \underset{\kappa}{\operatorname{argmax}}\, p(C|\mathcal{M}_\kappa) \quad , \tag{3.22}$$

which is called *maximum likelihood estimation (ML)*. Taking into account both classification and localization, the maximum a-posteriori estimation (MAP) looks like

$$(\widehat{\kappa}, \widehat{\boldsymbol{\phi}}, \widehat{\boldsymbol{t}}) = \underset{(\kappa,\boldsymbol{\phi},\boldsymbol{t})}{\operatorname{argmax}}\, p(\Omega_\kappa, \boldsymbol{\phi}, \boldsymbol{t}|C) = \underset{(\kappa,\boldsymbol{\phi},\boldsymbol{t})}{\operatorname{argmax}}\, p(\Omega_\kappa, \boldsymbol{\phi}, \boldsymbol{t}) p(C|\mathcal{M}_\kappa(\boldsymbol{\phi}, \boldsymbol{t})) \quad . \tag{3.23}$$

Assuming that the density value $p(\Omega_\kappa, \boldsymbol{\phi}, \boldsymbol{t})$ is the same for each object class Ω_κ and each pose parameters $(\boldsymbol{\phi}, \boldsymbol{t})$ the MAP estimation reduces to the ML estimation

$$(\widehat{\kappa}, \widehat{\boldsymbol{\phi}}, \widehat{\boldsymbol{t}}) = \underset{(\kappa,\boldsymbol{\phi},\boldsymbol{t})}{\operatorname{argmax}}\, p(C|\mathcal{M}_\kappa(\boldsymbol{\phi}, \boldsymbol{t})) \quad . \tag{3.24}$$

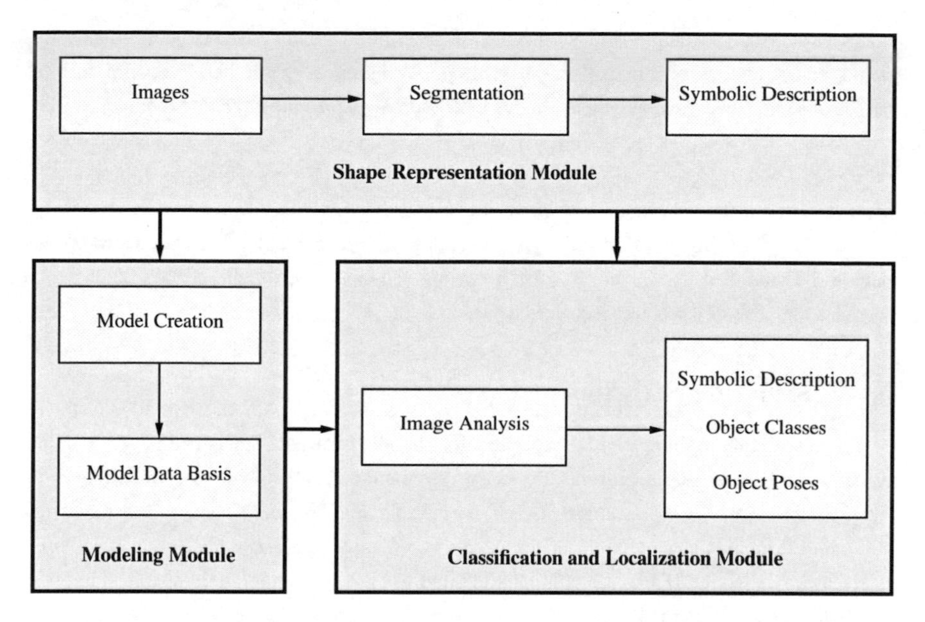

Figure 3.6: General scheme of a system for shape-based object classification and localization with all its fundamental modules [Win94]. The system consists of three modules, namely the shape representation module, the modeling module, and the classification and localization module.

In the present work, the statistical way is used. The density functions which are maximized in the recognition phase are created in the training phase. A definition of the density functions, which describe the objects in this work, follows in Chapter 4. More about the maximization algorithm can be found in Chapter 5.

3.2 Shape-Based Approaches

Winzen presents in [Win94] a general scheme of a system for shape-based[2] object classification and localization with all its fundamental modules (see Figure 3.6). The *shape representation module* (Section 3.2.1) is responsible for the initial symbolic description of the objects which is needed for both the *modeling module* (Section 3.2.2) and the *classification and localization mod-*

[2] Sometimes called also *model-based*.

ule (Section 3.2.3). Examples for such a symbolic description are image segmentation results like points, edges, regions, or corners [Pau92]. Sometimes relations between these geometric features can be observed, e. g., region neighborhoods or corner angles. In the modeling module, a model data basis which contains models $\mathcal{M} = \{\mathcal{M}_1, \mathcal{M}_2, \ldots, \mathcal{M}_\kappa, \ldots, \mathcal{M}_{N_\Omega}\}$ of all possible objects $\Omega = \{\Omega_1, \Omega_2, \ldots, \Omega_\kappa, \ldots, \Omega_{N_\Omega}\}$ used in the training phase of a particular recognition task is created. This model data basis is required in the classification and localization module for the comparison of training object features with features from a particular test scene in which objects have to be classified and localized. The following sections describe these three fundamental modules of shape-based object recognition systems.

3.2.1 Shape Representation

The representation of complex 3D shapes requires hundreds of parameters and is very complicated. There exist many approaches for shape representation which can be divided into three main groups, namely *surface-based*, *discontinuity-based*, and *volumetric-based*. A detailed discussion of these methods is not the objective of the present work; thus, only a short description is given in this section. Further details can be found in the literature references which are given for each representation method.

Surface-Based Representation

In this class of representations, surface properties, such as the surface normal and Gaussian curvature, are approximated and used in the description of the collected data points. Two main techniques in surface reconstruction exist, which may be used for surface-based segmentation and representation:

- *Gaussian Sphere [Hor86]*
  In this representation, the orientation of the surface normal at each point of the surface is mapped onto the corresponding point on the unit sphere. This representation is rotation invariant, because as the object is rotated, the mapped normals on the Gaussian sphere rotate accordingly. However, translation and the information about the dimensions of the objects are not preserved.

- *Moment-Based Representation [Tau89]*
  Shapes can be defined as functions of mass distribution, and their properties as moments of these functions. Using such a definition a so-called *shape polynomial* can be applied

to measure similarities between shapes. Similar to most moment independent representations, this representation suffers from sensitivity to occlusion.

Discontinuity-Based Representation

Rather than storing information about the surfaces, the discontinuity-based representation preserves the information about the points where the characteristic of the area on the surface changes. This results in curves embedded in 3D space, lowering storage requirements and increasing the efficiency in the higher level algorithms. The detected curves are represented by mathematical means such as parametric polynomial curves and rational B-splines. However, in most cases, the representation is not complete since information about the surface is lost. The discontinuity-based representation methods can be divided into:

- *Space Curve [Mok88]*
  Space curves are used to describe bounding contours and the spine of the sweep representations. Mokhtarian [Mok88] uses the general forms of expression for curvature and torsion to describe the curves over several scales.

- *Surface Primal Sketch [Pon87]*
  This representation, which was introduced by Ponce [Pon87], detects and models several types of discontinuities: *steps* - where the depth map is discontinuous, *roofs* - where the surface normal is discontinuous, *smooth joins* - where the principal curvature is discontinuous, and, finally, *shoulder* - a combination of two roofs. After Gaussian smoothing at a set of scales, the principal directions and curvature are computed everywhere for each smoothed image. Next, the zero crossings of the Gaussian curvature points are marked, and, finally, the descriptions of the four types of discontinuities are matched to detect the discontinuity points on the surface.

- *Aspect Graph [Koe79]*
  First proposed by Koenderik [Koe79], aspect graphs are graphs in which each node represents a distinct 2D viewpoint of a 3D object, and the arcs represent the transformations. Essentially, the graph is a partitioning of the viewing directions into stable regions. In a stable viewpoint, small changes in the viewing directions do not change the object aspect.

- *Discontinuity Labeling [God89]*
  Godin [God89] uses crease and jump edges to construct an edge junction graph. The nodes represent junctions, and the links symbolize the edges connecting the junctions.

Volumetric Representation

This class of representations describes volumes rather than surfaces or discontinuities of surfaces. While most such representation schemes are efficient in describing shapes, their disadvantage is that first, in most cases, the objects have to be symmetric and simple in shape. Second, these representations may not be used directly in the matching stage (classification and localization module), i. e., other properties must be derived from the representation. Approaches for volumetric representations can be divided into:

- *Superquadrics [Bar81]*
  This representation method was first introduced by Barr [Bar81]. Several parameters are used to determine the object squareness, size, bending, tapering, and twisting. Minimization techniques are used to recover the parameters from a set of data. However, since the minimization problem is nonlinear, good initial estimates are required for numerical stability. The major disadvantage of this representation is its lack of uniqueness, i. e., the same shape may be approximated by more than one set of parameters.

- *Octrees [Jac80]*
  Octrees, introduced by Jackins [Jac80], are a second class of volumetric representations. A tree structure, the octree, is constructed by recursively decomposing a cubical volume until each resultant subcube is homogeneous with respect to some criterion.

- *Constructive Solid Geometry [Req78]*
  Constructive solid geometry allows a modeler to create a complex surface or object by using Boolean operators to combine primitive solids. Often an object or surface that appears visually very complex is modeled as a superposition of cleverly combined primitive solids.

- *Sweep [Sou95]*
  In sweep representations, the desired shape is described by the sweeping action of a 2D function in the 3D space. The sweeping path is referred to as the spine, the function swept is referred to as the cross action, and the geometrical relationships of the cross section and the spine are referred to as the sweeping rule.

3.2.2 Modeling

The object symbolic descriptions, which are obtained with methods described in the previous section (Section 3.2.1), are used for both the model generation (modeling module) and the matching stage (classification and localization module, see Figure 3.6). In this section the model generation step (modeling module) of a shape-based object recognition system is presented.

Models are the a-priori geometrical and topological knowledge about objects. This knowledge is compared with the descriptions of the input data, obtained through lower level precesses, in the matching stage (classification and localization module). For 3D shape-based recognition systems, models must have knowledge of the shape of the object. There are two main approaches to model generation: sensor-based and through a *CAD/CAM*[3] *system*. In the sensor-based approach, multiple viewpoints of the object are integrated into a coherent fashion to provide a 3D description of the object. In the second approach, a CAD/CAM system is used, where a set of predefined primitives allows the user to construct interactively the CAD model of an object.

There exist two main procedures for model generation using the sensor-based approach: the 3D model of an object can be computed from a sequence of its 2D views (e. g., [Tom92, Wen93]), or 3D sensors are applied and no depth information has to be determined. However, it is not trivial to find the corresponding features in the different 2D views of an object, which is necessary for obtaining a correct object model.

There are several disadvantages of the sensor-based approaches versus the CAD/CAM systems for model generation. First, errors present in the data acquisition step, due to, e. g., calibration error, digitalization noise, and system distortions, can affect the modeling process. Second, techniques to register the data taken from all the 2D viewing directions must be developed and applied. On the other hand, there are many advantages in using a CAD/CAM system to model objects. First, it is automated, i. e., the object is designed interactively using the operations and the geometric primitives provided by the CAD/CAM system. Second, the system is more practical, and standardized modeling steps exist. However, there are also problems in using a CAD/CAM system. When designing a part, the designer uses a suitable representation for the part, and to manufacture the part, the necessary information is derived from the chosen representation. On the other hand, in object recognition, the reverse task must be performed, i. e., given the data points, a suitable representation must be derivable. The difficulty is that many entities used to design the object initially, such as the axis of symmetry, are not visible to the sensing module. Therefore, the vision system must first recognize the need for such entity, and second, the entity must be inferred. Nowadays the sensor-based way for object modeling rather than a CAD/CAM system is chosen for the development of shape-based object recognition systems.

[3]CAD/CAM - Computer Aided Design/Computer Aided Manufacturing

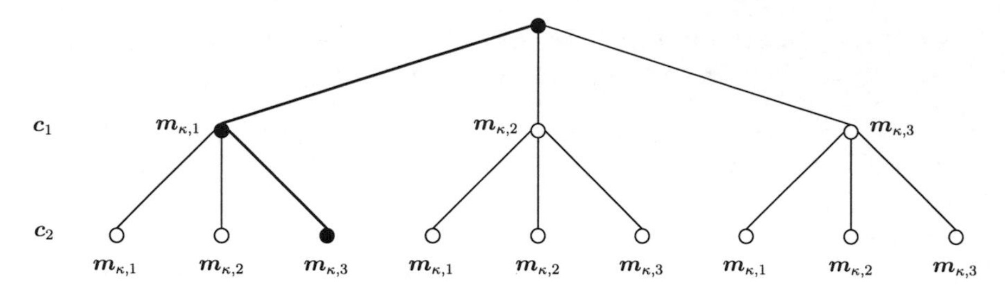

Figure 3.7: Correspondence problem for two scene features (c_1 and c_2) and three model features ($m_{\kappa,1}$, $m_{\kappa,2}$, and $m_{\kappa,3}$). In this example c_1 is assigned to $m_{\kappa,1}$ and c_2 corresponds to $m_{\kappa,3}$.

3.2.3 Classification and Localization

Once the model data basis with models $\mathcal{M} = \{\mathcal{M}_1, \mathcal{M}_2, \ldots, \mathcal{M}_\kappa, \ldots, \mathcal{M}_{N_\Omega}\}$ of all objects, which are taken into consideration for the recognition task, is created, the shape-based recognition system is able to classify and localize an object in an arbitrary test scene. First, a symbolic description of the object in the test scene is obtained using one of the methods described in Section 3.2.1, which results in a set of scene features c_i ($i = 1, 2, \ldots, N_C$). Then an algorithm for object classification and localization compares the scene features c_i with the model features $m_{\kappa,j}$ ($j = 1, 2, \ldots, M$) stored in the model data basis for each possible object class $\Omega = \{\Omega_1, \Omega_2, \ldots, \Omega_\kappa, \ldots, \Omega_{N_\Omega}\}$. Each of the scene features c_i is assigned to a model feature $m_{\kappa,j}$. For this purpose, an assignment function ζ is defined, where $\zeta_{\kappa,i} = j$ assigns the scene feature c_i to the model feature $m_{\kappa,j}$ of the object class Ω_κ [Hor96]. An illustration of this correspondence problem can be seen in Figure 3.7. Due to segmentation errors, the number of detected scene features N_C if often not equal to the number of model features M. Moreover, sometimes the segmentation results in the recognition phase are located not exactly on the same places as in the training phase. These and other problems make the searching for the best feature correspondence very difficult.

In order to evaluate feature correspondence, some authors define a measure of quality, which is maximized in the object recognition process, others introduce an error measure, which has to be minimized to find the class and pose of an object in a scene. The object classification and localization by error measure minimization is realized according to equation (3.19). Statistical algorithms for shape-based object recognition use the ML estimation defined in the equation (3.24).

3.3 Appearance-Based Approaches

In the present section, some known appearance-based algorithms for object recognition are discussed. The difference of these methods in comparison to the shape-based approaches lies in the feature extraction, which is presented in Section 3.3.1. Furthermore, some known terms like *template matching* (Section 3.3.2), *eigenspace approach* (Section 3.3.3), *neural networks* (Section 3.3.4), and *support vector machines* (Section 3.3.5) are also shortly described.

3.3.1 Feature Vectors

In contrast to shape-based approaches, appearance-based methods do not use any previous segmentation steps in order to extract object features. The feature vectors are computed directly from the pixel values in this case. Depending on the chosen way a global feature vector $\boldsymbol{c}$ or a set of local feature vectors C (3.14) is extracted from the image using appropriate mathematical operation, which was introduced in Section 3.1.4.

For computation of a global feature vector $\boldsymbol{c}$, either all pixels $\boldsymbol{f}_{x,y}$ from the whole image $\boldsymbol{f}$ or all pixels $\boldsymbol{f}_{x,y}$ within an object area (region of interest) O_κ are taken into consideration. They are usually transformed to an orthogonal basis $\boldsymbol{\Phi}$

$$\boldsymbol{c} = \boldsymbol{\Phi}\boldsymbol{f}' \quad , \tag{3.25}$$

whereas $\boldsymbol{f}'$ denotes the image $\boldsymbol{f}$ or its part written as a vector. In the case of gray level images, the pixels are represented by scalars $f_{x,y}$ and the vector $\boldsymbol{f}'$ can be written as

$$\boldsymbol{f}' = (f_{1,1}, \ldots, f_{1,N_y}, \ldots, f_{N_x,1}, \ldots, f_{N_x,N_y})^{\mathrm{T}} \quad . \tag{3.26}$$

An example for such a feature extraction is the eigenspace approach (e. g., [Tur91, Grä03]), of which the description follows in Section 3.3.3. There exist also object recognition systems computing a global feature vector $\boldsymbol{c}$ using the wavelet transformation presented in Section 2.2. Some authors use the standard definition of this transformation [Pap00, Sco00]. Others apply modified wavelets to describe objects [Mal97].

Local feature vectors $\boldsymbol{c}_m$ are computed for small parts of images. Some authors, e. g., [Sho96, Low99], extract the local feature vectors only on the so-called keypoints, which can be determined with the *Harris detector* [Har88]. However, usually images are divided into squares of size $\Delta r = \Delta x = \Delta y$ and local feature vectors $\boldsymbol{c}_m$ are computed for all of these squares in their centers (grid points) $\boldsymbol{x}_m$, i. e., $\boldsymbol{c}_m = \boldsymbol{c}(\boldsymbol{x}_m)$, which is illustrated in Figure 3.8. For local

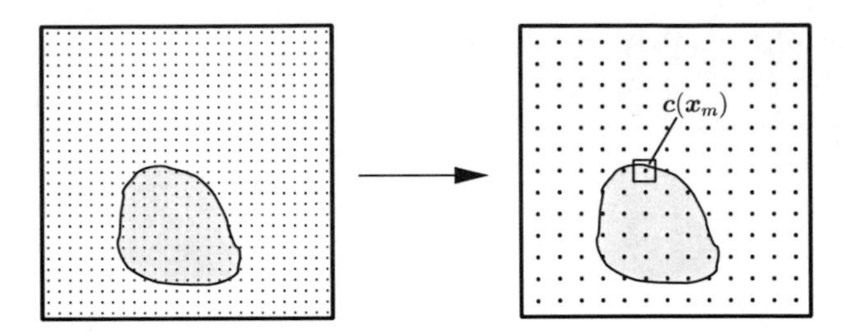

Figure 3.8: Left: an image $\boldsymbol{f}$ with pixels $\boldsymbol{f}_{x,y}$. Right: a grid is laid on the image $\boldsymbol{f}$ and a local feature vector $\boldsymbol{c}_m = \boldsymbol{c}(\boldsymbol{x}_m)$ is computed on each grid point.

feature extraction, the same transformations as for global feature vectors can be used. They are performed locally in this case. In [dV98] the eigenspace approach, sometimes also called *Principal Component Analysis (PCA)*, is applied locally for neighborhoods of size 9×9 pixels. In many cases results of the wavelet transformation (Section 2.2) are taken into account for the local feature extraction. Examples can be found in [Pös99, Rei05, Grz05a]. Furthermore, local filters like:

- *Derivative of the Gaussian Distribution [Sch99]*
- *Gabor Filter [Wal00]*
- *Gradient Filter [Hua06]*
- *Laplace Operator [Bla06]*

are also often applied to local feature extraction in object recognition systems.

3.3.2 Template Matching

Template matching is one of the most fundamental approaches for appearance-based object classification and localization [Bru97, Gon01, Pra01]. According to the general concept of 3D object recognition described in Section 1.1, images of all objects Ω_κ are acquired in the training phase from different viewpoints $(\boldsymbol{\phi}_{\text{ext}}, \boldsymbol{t}_{\text{ext}})$ (see Figure 1.3). Object modeling, which results in models $\mathcal{M}_\kappa$ for all object classes Ω_κ, is not performed in this case. Instead, rectangular windows containing the objects are cut out from the training images and stored as so called templates $\boldsymbol{T}_\kappa(\boldsymbol{\phi}_{\text{ext}}, \boldsymbol{t}_{\text{ext}})$. The templates for each object class Ω_κ are written as functions of external pose

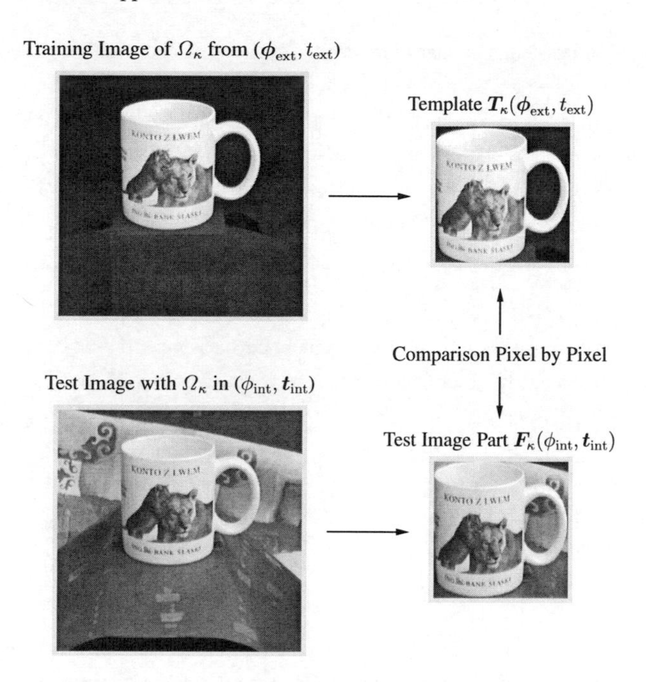

Figure 3.9: First row: training image of the object Ω_κ taken from the view $(\boldsymbol{\phi}_{\text{ext}}, t_{\text{ext}})$ and the corresponding template $\boldsymbol{T}_\kappa(\boldsymbol{\phi}_{\text{ext}}, t_{\text{ext}})$. Second row: test image with the object Ω_κ located in the pose $(\phi_{\text{int}}, \boldsymbol{t}_{\text{int}})$ on heterogeneous background and one of its parts $\boldsymbol{F}_\kappa(\phi_{\text{int}}, \boldsymbol{t}_{\text{int}})$, which is compared with the template in the recognition phase.

parameters $(\boldsymbol{\phi}_{\text{ext}}, t_{\text{ext}})$, because for each training view a different template is stored. In the recognition phase, all these templates $\boldsymbol{T}_\kappa(\boldsymbol{\phi}_{\text{ext}}, t_{\text{ext}})$ are translated and rotated with the internal transformations $(\phi_{\text{int}}, \boldsymbol{t}_{\text{int}})$ over the whole test image. They are compared pixel by pixel with the corresponding parts $\boldsymbol{F}_\kappa(\phi_{\text{int}}, \boldsymbol{t}_{\text{int}})$ of the test image. The corresponding parts of the test image for each searched object class Ω_κ are written as functions of internal pose parameters $(\phi_{\text{int}}, \boldsymbol{t}_{\text{int}})$, which denote the current pose of the templates $\boldsymbol{T}_\kappa(\boldsymbol{\phi}_{\text{ext}}, t_{\text{ext}})$ in the test image. An example training image, corresponding template $\boldsymbol{T}_\kappa(\boldsymbol{\phi}_{\text{ext}}, t_{\text{ext}})$, test image, and one of its parts $\boldsymbol{F}_\kappa(\phi_{\text{int}}, \boldsymbol{t}_{\text{int}})$, which is compared with the template in the recognition phase, are depicted in Figure 3.9. In this example, the parameters $(\kappa, \phi_{\text{int}}, \boldsymbol{\phi}_{\text{ext}}, \boldsymbol{t}_{\text{int}}, t_{\text{ext}})$, which can be written together as $(\kappa, \boldsymbol{\phi}, \boldsymbol{t})$, are knowingly chosen to be correct regarding the classification and localization result. In this way, it can be shown that even such similar template $\boldsymbol{T}_\kappa(\boldsymbol{\phi}_{\text{ext}}, t_{\text{ext}})$ and test image part $\boldsymbol{F}_\kappa(\phi_{\text{int}}, \boldsymbol{t}_{\text{int}})$

do not yield a perfect template matching result. In this case, the template matching algorithm compares also pixels which belong to the background and not to the object. For test scenes with unknown heterogeneous background, like in Figure 3.9, such a comparison yields a very big difference for these pixels, and a successful classification and localization is not possible. Since a template match is rarely exact, a common procedure is to compute a difference measure between the template $\boldsymbol{T}_\kappa(\boldsymbol{\phi}_{\text{ext}}, t_{\text{ext}})$ and the test image part $\boldsymbol{F}_\kappa(\phi_{\text{int}}, \boldsymbol{t}_{\text{int}})$

$$d(\boldsymbol{T}_\kappa(\boldsymbol{\phi}_{\text{ext}}, t_{\text{ext}}), \boldsymbol{F}_\kappa(\phi_{\text{int}}, \boldsymbol{t}_{\text{int}})) = d(\kappa, \boldsymbol{\phi}, \boldsymbol{t}) \quad . \tag{3.27}$$

Independent on the exact definition of this difference measure d the object classification and localization problem can be described according to equation 3.19 with

$$(\widehat{\kappa}, \widehat{\boldsymbol{\phi}}, \widehat{\boldsymbol{t}}) = \underset{(\kappa, \boldsymbol{\phi}, \boldsymbol{t})}{\operatorname{argmin}}\, d(\kappa, \boldsymbol{\phi}, \boldsymbol{t}) \quad , \tag{3.28}$$

or more detailed with

$$(\widehat{\kappa}, \widehat{\phi}_{\text{int}}, \widehat{\boldsymbol{\phi}}_{\text{ext}}, \widehat{\boldsymbol{t}}_{\text{int}}, \widehat{t}_{\text{ext}}) = \underset{(\kappa, \phi_{\text{int}}, \boldsymbol{\phi}_{\text{ext}}, \boldsymbol{t}_{\text{int}}, t_{\text{ext}})}{\operatorname{argmin}} d(\boldsymbol{T}_\kappa(\boldsymbol{\phi}_{\text{ext}}, t_{\text{ext}}), \boldsymbol{F}_\kappa(\phi_{\text{int}}, \boldsymbol{t}_{\text{int}})) \quad . \tag{3.29}$$

As difference measure d any function satisfying conditions of a distance function can be used. Usually correlations between the templates $\boldsymbol{T}_\kappa(\boldsymbol{\phi}_{\text{ext}}, t_{\text{ext}})$ and the image parts $\boldsymbol{F}_\kappa(\phi_{\text{int}}, \boldsymbol{t}_{\text{int}})$ are computed and minimized in order to determine the class $\widehat{\kappa}$ and the pose $(\widehat{\boldsymbol{\phi}}, \widehat{\boldsymbol{t}})$ of an object in the image. This simple approach for object recognition suffers from disadvantages like:

- *Pixels as Features*
  Usually, in the template matching algorithm, the templates $\boldsymbol{T}_\kappa(\boldsymbol{\phi}_{\text{ext}}, t_{\text{ext}})$ are compared with the test image parts $\boldsymbol{F}_\kappa(\phi_{\text{int}}, \boldsymbol{t}_{\text{int}})$ pixel by pixel, i. e., pixel values are used as features. This solution is very sensitive to noise and yields bad results in a real world environment.

- *Discrete External Pose Parameter Space*
  As localization results only these external pose parameters can be obtained, which correspond to the training views. However, in the recognition phase the objects can lie between the training views. In this case, an exact object localization is not possible using the template matching approach.

- *Large Data Amount*
  In the training phase, templates for all objects from all training views are stored. Object feature extraction and modeling, which reduces the amount of data, is not performed.

- *Internal Rotation*

  If in the recognition phase the object is rotated in the image plane, i. e., with the internal rotation ϕ_{int}, the template matching algorithm has to deal with an additional problem. The pixels from the templates $\boldsymbol{T}_\kappa(\boldsymbol{\phi}_{\text{ext}}, \boldsymbol{t}_{\text{ext}})$ do not lie exactly on the same places as the pixels from the corresponding image parts $\boldsymbol{F}_\kappa(\phi_{\text{int}}, \boldsymbol{t}_{\text{int}})$. In order to compute the difference measure $d(\kappa, \boldsymbol{\phi}, \boldsymbol{t})$, a pixel interpolation has to be performed, which additionally increases the noise and the difference value.

Due to these disadvantages, other, more complex algorithms and methods for appearance-based object recognition were developed. The most relevant ones are described in the following sections.

3.3.3 Eigenspace Approach

The eigenspace approach [Tur91, Leo96, Mog97, Grä03], sometimes also called Principal Component Analysis (PCA), which is based on the *Karhunen-Loeve Transform* [Kar46, Loe55], usually describes objects with global feature vectors (3.15). Let us assume we have N_ρ training images and N_ρ image vectors $\boldsymbol{f}'_{\rho=1,\ldots,N_\rho}$ for them according to (3.26). As mentioned in Section 3.3.1, a global feature vector $\boldsymbol{c}_\rho$ is computed for the training image vector $\boldsymbol{f}'_\rho$ using an orthonormal matrix $\boldsymbol{\Phi}$

$$\boldsymbol{c}_\rho = \boldsymbol{\Phi}(\boldsymbol{f}'_\rho - \boldsymbol{\mu}_{\boldsymbol{f}'}) \quad , \tag{3.30}$$

where $\boldsymbol{\mu}_{\boldsymbol{f}'}$ is a mean value vector determined for all N_ρ training images. The question is how the Principal Component Analysis determines the mean value vector $\boldsymbol{\mu}_{\boldsymbol{f}'}$ and the transformation matrix $\boldsymbol{\Phi}$ for a given data set. For this purpose, some fundamental statistical terms introduced in Section 2.1 are used. First, the image vector (3.26) is considered as a random vector (2.1)

$$\begin{aligned} \boldsymbol{f}' = \Big(\; & f_{1,1}, f_{1,2}, \ldots, f_{1,N_y}, \\ & f_{2,1}, f_{2,2}, \ldots, f_{2,N_y}, \\ & \vdots \\ & f_{N_x,1}, f_{N_x,2}, \ldots, f_{N_x,N_y} \; \Big)^{\mathrm{T}} \end{aligned} \quad . \tag{3.31}$$

The elements of this vector are uniformly distributed (2.22) random variables (Definition 2.1) with N_ρ possible values

$$f_{x,y} \in \{f_{1,x,y}, \ldots, f_{\rho,x,y}, \ldots, f_{N_\rho,x,y}\} \quad , \tag{3.32}$$

which correspond to the N_ρ training image vectors. Since the random variables $f_{x,y}$ are uniformly distrubuted their mean values $\mu_{x,y}$ (Definition 2.4) can be computed with

$$\mu_{x,y} = \frac{1}{N_\rho} \sum_{\rho=1}^{N_\rho} f_{\rho,x,y} \quad . \tag{3.33}$$

In this way, the mean value vector $\boldsymbol{\mu}_{\boldsymbol{f}'}$, which is needed for equation (3.30), can be written as

$$\boldsymbol{\mu}_{\boldsymbol{f}'} = \frac{1}{N_\rho} \sum_{\rho=1}^{N_\rho} \boldsymbol{f}'_\rho \quad . \tag{3.34}$$

After the mean value vector $\boldsymbol{\mu}_{\boldsymbol{f}'}$ is determined, the Principal Component Analysis computes the transformation matrix $\boldsymbol{\Phi}$. First, a covariance matrix (2.16) for the image random vector $\boldsymbol{f}'$ (3.31) is calculated

$$\boldsymbol{\Sigma} = \begin{pmatrix} \sigma^2_{1,1} & \sigma^2_{1,2} & \cdots & \sigma^2_{1,N_x \cdot N_y} \\ \sigma^2_{2,1} & \sigma^2_{2,2} & \cdots & \sigma^2_{2,N_x \cdot N_y} \\ \vdots & \vdots & & \vdots \\ \sigma^2_{N_x \cdot N_y,1} & \sigma_{N_x \cdot N_y,2} & \cdots & \sigma^2_{N_x \cdot N_y, N_x \cdot N_y} \end{pmatrix} , \tag{3.35}$$

where the covariance between the $f_{x1,y1}$ and $f_{x2,y2}$ random variables under assumption of their uniform distributions can be expressed as

$$\text{cov}(f_{x1,y1}, f_{x2,y2}) = \sigma^2_{x1 \cdot y1, x2 \cdot y2} = \frac{1}{N_\rho} \sum_{\rho=1}^{N_\rho} (f_{\rho,x1,y1} - \mu_{x1,y1})(f_{\rho,x2,y2} - \mu_{x2,y2}) \quad . \tag{3.36}$$

Since the covariance matrix (3.35) is square, *eigenvectors* $\boldsymbol{u}_j$ and *eigenvalues* λ_j for this matrix can be calculated. Each vector $\boldsymbol{u}_j$ is called eigenvector of the matrix $\boldsymbol{\Sigma}$ and each scalar λ_j its corresponding eigenvalue, if

$$\boldsymbol{\Sigma} \boldsymbol{u}_j = \lambda_j \boldsymbol{u}_j \quad . \tag{3.37}$$

The methods for computation of the eigenvectors and their eigenvalues can be found in [Bro85]. Here it is assumed that $\boldsymbol{u}_{j=1,\ldots,N_x \cdot N_y}$ are eigenvectors of the covariance matrix $\boldsymbol{\Sigma}$ (3.35) sorted by the decreasing eigenvalues $\lambda_{j=1,\ldots,N_x \cdot N_y}$. They build the transformation matrix $\boldsymbol{\Phi}$ as follows

$$\boldsymbol{\Phi} = \begin{pmatrix} \boldsymbol{u}_1{}^{\mathrm{T}} \\ \boldsymbol{u}_2{}^{\mathrm{T}} \\ \vdots \\ \boldsymbol{u}_{N_x \cdot N_y}{}^{\mathrm{T}} \end{pmatrix} = \begin{pmatrix} u_{1,1} & u_{1,2} & \cdots & u_{1,N_x \cdot N_y} \\ u_{2,1} & u_{2,2} & \cdots & u_{2,N_x \cdot N_y} \\ \vdots & \vdots & & \vdots \\ u_{N_x \cdot N_y,1} & u_{N_x \cdot N_y,2} & \cdots & u_{N_x \cdot N_y, N_x \cdot N_y} \end{pmatrix} . \tag{3.38}$$

Using the formulas (3.30), (3.34), and (3.38) a global vector $\boldsymbol{c}_\rho$ for each training image vector $\boldsymbol{f}'_\rho$ can be computed. Directly from equation (3.30) it follows that the image vector $\boldsymbol{f}'_\rho$ can be reconstructed from the global feature vector $\boldsymbol{c}_\rho$ using

$$\boldsymbol{f}'_\rho = \boldsymbol{\Phi}^{\mathrm{T}} \boldsymbol{c}_\rho + \boldsymbol{\mu}_{\boldsymbol{f}'} \quad . \tag{3.39}$$

Since the most important information about the image $\boldsymbol{f}'_\rho$ is contained in the first eigenvectors $\boldsymbol{u}_{j=1,\ldots,N_E}$, the dimension of the feature vector $\boldsymbol{c}_\rho$ can be reduced from $N_x \cdot N_y$ to N_E ($N_E < N_x \cdot N_y$) with relatively little loss of information. The last $N_x \cdot N_y - N_E$ elements of the feature vector $\boldsymbol{c}_\rho$ are just not taken into account. In such a case, only an approximated image reconstruction is possible

$$\widehat{\boldsymbol{f}}'_\rho = \begin{pmatrix} u_{1,1} & u_{2,1} & \cdots & u_{N_E,1} \\ u_{1,2} & u_{2,2} & \cdots & u_{N_E,2} \\ \vdots & \vdots & & \vdots \\ u_{1,N_x \cdot N_y} & u_{2,N_x \cdot N_y} & \cdots & u_{N_E,N_x \cdot N_y} \end{pmatrix} \begin{pmatrix} c_{1,\rho} \\ c_{2,\rho} \\ \vdots \\ c_{N_E,\rho} \end{pmatrix} + \begin{pmatrix} \mu_{1,\boldsymbol{f}'} \\ \mu_{2,\boldsymbol{f}'} \\ \vdots \\ \mu_{N_E,\boldsymbol{f}'} \end{pmatrix} , \tag{3.40}$$

where $\widehat{\boldsymbol{f}}'_\rho$ is the approximated image vector. Note that the dimension of the mean value vector is also reduced. From this point until the end of Section 3.3.3, the transformation matrix $\boldsymbol{\Phi}$, the feature vector $\boldsymbol{c}_\rho$, and the mean value vector $\boldsymbol{\mu}_{\boldsymbol{f}'}$ are considered with the reduced first dimension to N_E as in the equation (3.40).

As mentioned, in the classical object classification and localization problem, a set of object classes $\Omega = \{\Omega_1, \Omega_2, \ldots, \Omega_\kappa, \ldots, \Omega_{N_\Omega}\}$ is considered. For all of these objects Ω_κ training images written as vectors $\boldsymbol{f}'_\rho$ are taken from different poses $(\boldsymbol{\phi}, \boldsymbol{t})$. For these training image vectors global features $\boldsymbol{c}_\rho$ using the Principal Component Analysis are computed. Besides the global feature vector $\boldsymbol{c}_\rho$, the class number κ and the pose $(\boldsymbol{\phi}, \boldsymbol{t})$ of the object in the training image $\boldsymbol{f}'_\rho$ have to be preserved in the object model. The global feature vector $\boldsymbol{c}_\rho$ with these additional informations $(\kappa, \boldsymbol{\phi}, \boldsymbol{t})$ is written together as a model vector $\boldsymbol{m}_\kappa(\boldsymbol{\phi}, \boldsymbol{t})$. The number of model

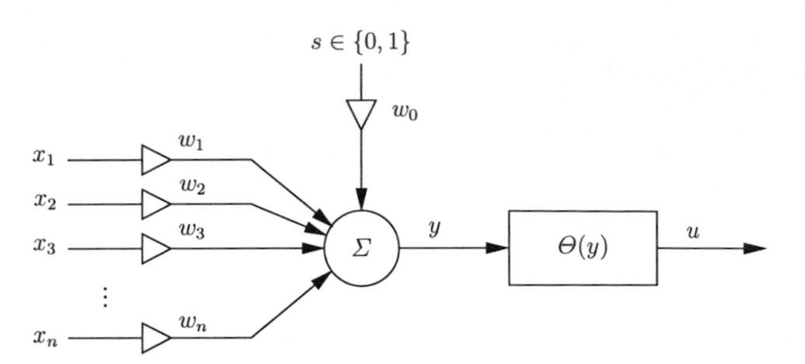

Figure 3.10: Artificial neuron with input signals x_i, weights w_i, an offset w_0, which can be set with s, a sum signal y, an activity function Θ, and an output signal u.

vectors is equal to the number of training images (feature vectors) N_ρ. As one can suppose, the algorithm for object classification and localization can be expressed as a minimization of a difference measure $d(\boldsymbol{c}, \boldsymbol{m}_\kappa(\boldsymbol{\phi}, \boldsymbol{t}))$

$$(\widehat{\kappa}, \widehat{\boldsymbol{\phi}}, \widehat{\boldsymbol{t}}) = \underset{(\kappa,\boldsymbol{\phi},\boldsymbol{t})}{\operatorname{argmin}}\, d(\boldsymbol{c}, \boldsymbol{m}_\kappa(\boldsymbol{\phi}, \boldsymbol{t})) \quad , \tag{3.41}$$

where $\boldsymbol{c}$ is global feature vector computed from the test image vector, in which an object has to be classified and localized.

3.3.4 Neural Networks

Artificial neural networks are able to approximate any arbitrary function. For this reason, they are excellent candidates for realization of systems for object classification and localization [Nie83]. An artificial neural network (ANN), also called a simulated neural network (SNN) or just a neural network (NN), is an interconnected group of *artificial neurons* [Rip96, Tad98, Tad99, Wal00, Yua01, Par04]. An artificial neuron is a device with many inputs x_i and one output u (see Figure 3.10). The inputs x_i are weighted with the weights w_i. The weighted input signals and, if necessary ($s = 1$), the offset w_0 are added together

$$y = s w_0 + \sum_{i=1}^{n} w_i x_i \quad . \tag{3.42}$$

The sum y is then processed by an activity function $\Theta(y)$, which results in the neuron output u. As activity function, a nonlinear function like

$$\Theta(y) = \frac{1}{1 + \exp(-\beta y)} \tag{3.43}$$

can be used. Using equations (3.42) and (3.43), the output signal u is given by

$$u = \frac{1}{1 + \exp[-\beta(sw_0 + \sum_{i=1}^{n} w_i x_i)]} \tag{3.44}$$

The number n of the input signals, the weight values w_i, and the coefficient β can be different for different neurons in a neural network. There are many possibilities to connect the nodes (artificial neurons) in a neural network. The most widely used type of neural network is called *multilayer perceptron* [Gon01]. Input signals x_i are processed through successive layers of neurons. There is always an input and an output layer. The layers between them are called hidden layers. All problems which can be solved by a multilayer perceptron can be solved with only one hidden layer, but it is sometimes more efficient to use two or more hidden layers. Once the neural network topology, which defines the nodes and the connections between them, is determined, it does not change anymore. The weights $w_{i=1,\ldots,n}$, the coefficient β, and the offset w_0 are trained for each neuron using a sample set of images. A detailed overview of object recognition methods which use neural networks can be found in [EP02].

3.3.5 Support Vector Machines

Support Vector Machines (SVM) have been proposed as a very effective method for general purpose pattern recognition [Cor95, Vap95]. Intuitively, given a set of points which belong to either of two classes, a Support Vector Machine finds the hyperplane leaving the largest possible fraction of points of the same class on the same side, while maximizing the distance of either class from the hyperplane. In the sense of object classification in digital images, a simple two-class problem has to be solved for all objects $\Omega_{\kappa=1,\ldots,N_\Omega}$ considered in the task. The first class is the object class Ω_κ itself. The second class represents everything which is not the class Ω_κ, and can be denoted by Ω'_κ. For the training of the class Ω_κ images of this object Ω_κ from different viewpoints, while for the learning of the anti-class Ω'_κ images of all other objects $\Omega_{i\neq\kappa}$, are taken into account. In the recognition phase, the Support Vector Machine decides which of the objects $\Omega_{\kappa=1,\ldots,N_\Omega}$ occurs in a test scene. The two-class problem is regarded for each object class Ω_κ,

i. e., N_Ω times. It is expected that $N_\Omega - 1$ times the anti-class $\Omega'_{\widehat{\kappa}}$, and exactly once the class $\Omega_{\widehat{\kappa}}$, wins, where $\widehat{\kappa}$ is the classification result[4].

In the following a simple case, where the object class Ω_κ and its anti-class Ω'_κ are linearly separable, is discussed. Let the feature vectors $\boldsymbol{c}_{j=1,\dots,N_S}$ representing all object classes $\Omega_{\kappa=1,\dots,\kappa}$ build a set S

$$S = \{\boldsymbol{c}_1, \boldsymbol{c}_2, \dots, \boldsymbol{c}_j, \dots, \boldsymbol{c}_{N_S}\} \quad . \tag{3.45}$$

Note that the set C (3.14) defined in Section 3.1.4 is not equal to S, because C contains feature vectors describing only one object. Each feature vector $\boldsymbol{c}_j$ from S belongs either to the class Ω_κ or to the anti-class Ω'_κ, which is given with the corresponding labels $y_j = \{-1, 1\}$. The goal is to establish the equation of a hyperplane that divides the set S leaving all the feature vectors describing Ω_κ on its one side and all the feature vectors belonging to Ω'_κ on the other side of the hyperplane. Moreover, both the distance of the class Ω_κ and the anti-class Ω'_κ to the hyperplane has to be maximized. For this purpose, some preliminary definitions are needed.

Definition 3.1 *The set S is linearly separable if there exist a vector $\boldsymbol{v} \in \mathbb{R}^{N_c}$ and scalar $b \in \mathbb{R}$ such that*

$$y_j(\boldsymbol{v} \cdot \boldsymbol{c}_j + b) \geq 1 \tag{3.46}$$

for all $j = 1, 2, \dots, N_S$. Note that $\boldsymbol{c}_j \in \mathbb{R}^{N_c}$ (3.16).

The pair $(\boldsymbol{v}, b)$ defines a hyperplane of equation

$$\boldsymbol{v} \cdot \boldsymbol{c} + b = 0 \quad , \tag{3.47}$$

named *separating hyperplane*. If with $|\boldsymbol{v}|$ the norm of the vector $\boldsymbol{v}$ is denoted, the distance d_j of a point $\boldsymbol{c}_j$ (feature vector) to the separating hyperplane $(\boldsymbol{v}, b)$ is given by

$$d_j = \frac{\boldsymbol{v} \cdot \boldsymbol{c}_j + b}{|\boldsymbol{v}|} \quad . \tag{3.48}$$

Combining inequality (3.46) and equation (3.48) for all $\boldsymbol{c}_j \in S$ we have

$$y_j d_j \geq \frac{1}{|\boldsymbol{v}|} \quad . \tag{3.49}$$

Therefore, $|\boldsymbol{v}|^{-1}$ is the lower bound on the distance between the feature vectors $\boldsymbol{c}_j$ and the separating hyperplane $(\boldsymbol{v}, b)$. Now a *canonical representation* of the separating hyperplane is ob-

[4] Assuming that exactly one of the trained objects $\Omega_{\kappa=1,\dots,N_\Omega}$ occurs in the scene.

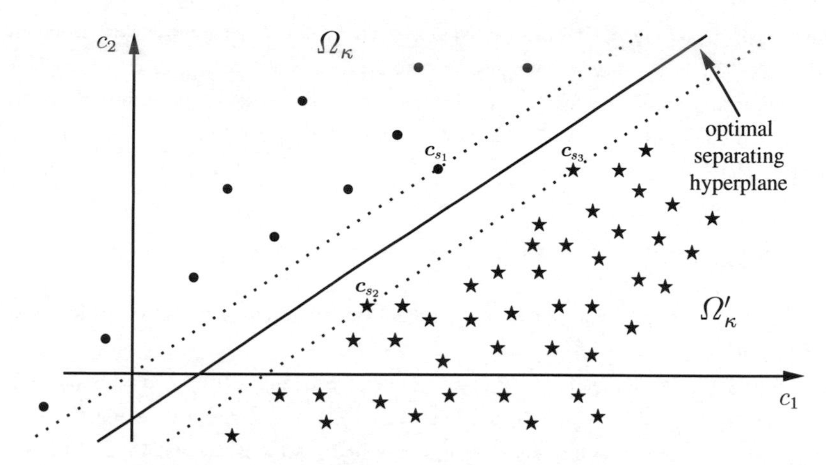

Figure 3.11: Optimal separating hyperplane for two-dimensional feature space. With • feature vectors of the object class Ω_κ are denoted. By ★ the remaining feature vectors of all other object classes are represented. Three feature vectors $\boldsymbol{c}_{s_1}$, $\boldsymbol{c}_{s_2}$, and $\boldsymbol{c}_{s_3}$ lie in the minimum distance to the optimal separating hyperplane and are called support vectors.

tained by rescaling the pair $(\boldsymbol{v}, b)$ into the pair $(\boldsymbol{v}', b')$ in such a way that the distance of the closest feature vector equals $|\boldsymbol{v}'|^{-1}$. For the canonical representation $(\boldsymbol{v}', b')$ of the hyperplane considering the equation (3.46) it can be written

$$\min_{\boldsymbol{c}_j \in S}\{y_j(\boldsymbol{v}' \cdot \boldsymbol{c}_j + b')\} = 1 \quad . \tag{3.50}$$

Consequently, for a separating hyperplane in the canonical representation, the bound in inequality (3.49) is tight. The discussion comes to the point where the *optimal separating hyperplane* has to be defined.

Definition 3.2 *Given a linearly separable set S, the optimal separating hyperplane is the separating hyperplane, for which the distance to the closest point of S is maximum.*

Such an optimal separating hyperplane for two-dimensional feature space, i. e., for $\boldsymbol{c} = (c_1, c_2)^{\mathrm{T}}$, is depicted in Figure 3.11. In this case it is just a straight line. The feature vectors $\boldsymbol{c}_{s_1}$, $\boldsymbol{c}_{s_2}$, and $\boldsymbol{c}_{s_3}$, which are closest to the optimal separating hyperplane, are called *support vectors*. For object modeling (training phase) it is enough to store the support vectors for each class Ω_κ, which significantly reduces the data amount. In the recognition phase, the classification algorithm starts with the extraction of feature vectors from a scene. After that it determines for each object class, on which side of the optimal separating hyperplane the feature vectors lie. In this way, the objects which occur in the scene are found. The object localization is not possible with this algorithm. A detailed discussion of object classification methods using the Support Vector Machine can be found in [Chr00].

3.4 Summary

In this chapter, some common algorithms and methods for object classification and localization were presented.

Section 3.1 discussed fundamental terms of object recognition. It started with the description of the image acquisition process (Section 3.1.1), where the perspective, orthographic, weak perspective, and paraperspective camera projection models were discussed. General properties of image databases for object recognition tasks were analyzed in Section 3.1.2. As examples of such databases, the Columbia Object Image Library (COIL) and image database 3D object recognition (DIROKOL) were presented. According to the general scheme of an object recognition system depicted in Figure 1.3, before images are used for object feature extraction, they are first preprocessed. This additional preprocessing step was shortly introduced in Section 3.1.3. It was explained how to convert true color images to gray level images, how to resize them, and finally, an overview of methods for image segmentation was given. Section 3.1.4 dealt with object feature extraction, while Section 3.1.5 concerned object modeling and recognition.

The shape-based approaches for object classification and localization were considered in Section 3.2. A system for shape-based object recognition consists of three modules (see Figure 3.6), namely the shape representation module (Section 3.2.1), the modeling module (Section 3.2.2), and the classification and localization module (Section 3.2.3). The shapes can be represented using surface-based, discontinuity-based, and volumetric representation, which was described in Section 3.2.1. A comparison of the sensor-based approaches and CAD/CAM systems for model generation was given in Section 3.2.2. Section 3.2.3 described shortly the algorithm for object classification and localization in the case of shape-based methods.

Section 3.3 started with the description of object feature extraction (Section 3.3.1) in the case of appearance-based algorithms for object recognition. Then it presented some known approaches for appearance-based object classification and localization like template matching (Section 3.3.2), eigenspace approach (Section 3.3.3), neural networks (Section 3.3.4), and support vector machines (Section 3.3.5).

Chapter 4

Training of Statistical Object Models

As can be seen in Figure 1.3, a system for object classification and localization works in two different modes. The first one is the training[1] phase, the second one is called recognition. Before objects can be classified and localized in the recognition phase, the training of object models $\mathcal{M} = \{\mathcal{M}_1, \mathcal{M}_2, \ldots, \mathcal{M}_\kappa, \ldots, \mathcal{M}_{N_\Omega}\}$ for all object classes $\Omega = \{\Omega_1, \Omega_2, \ldots, \Omega_\kappa, \ldots, \Omega_{N_\Omega}\}$ considered in the particular task has to be performed. Moreover, the context dependencies between the objects $\Omega = \{\Omega_1, \Omega_2, \ldots, \Omega_\kappa, \ldots, \Omega_{N_\Omega}\}$ are learned in the training phase in order to improve the classification rates for multi-object scenes. In this chapter all steps of the learning mode for the object recognition system, which was taken over from [Rei04] and widely extended within the scope of the present work, are discussed.

Figure 4.1 illustrates step by step the training process for the system. The system starts the training with the image acquisition. It can be done in two different ways, namely with a turntable and camera arm, or with a hand-held camera. Using the turntable and camera arm, not only the images are acquired, but also the poses of the objects in the images are known. If the objects are taken with the hand-held camera, an additional step, namely pose parameter reconstruction, is introduced for object modeling, because the object poses are not known after the image acquisition in this way. The context modeling does not require the pose parameter reconstruction. These steps are depicted in Figure 4.1 with boxes marked with ■. They deliver the training data (training images, object poses) and are described in Section 4.1.

Section 4.2 presents the creation of a statistical object model using gray level images. After converting and resizing the original training images into square gray level images, two-dimensional local feature vectors are computed. These steps can be seen in Figure 4.1 in boxes marked with ★. Then the decision is made, which of the feature vectors describe the object, and

[1] Sometimes called also learning.

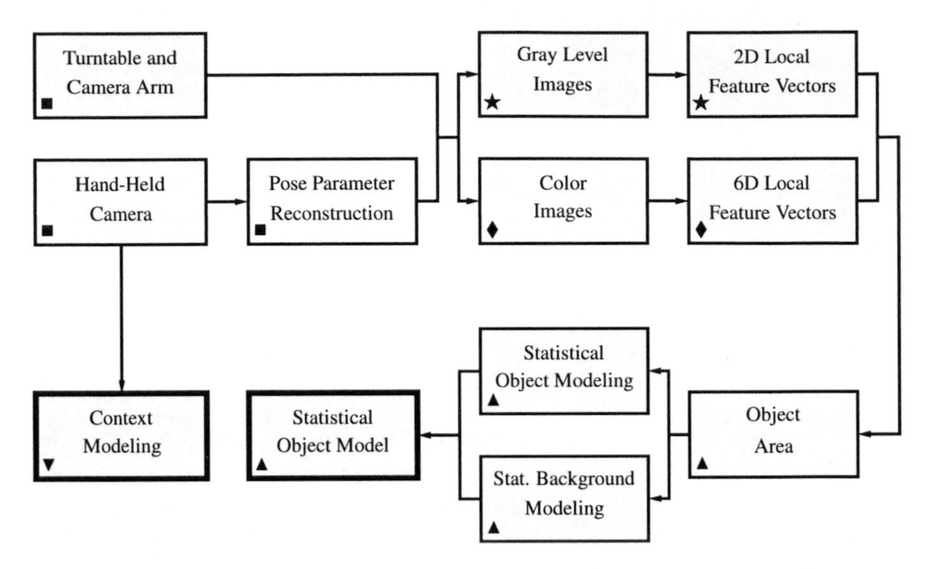

Figure 4.1: General scheme of the training phase. After the image acquisition and, if needed, the pose parameter reconstruction, the training images are converted either into gray level or color images. Then 2D or 6D feature vectors (respectively) are computed and the object area is defined. Finally, statistical object, background, and, if considered, context modeling are performed.

which belong to the background. For this purpose a so-called object area is defined. Finally, the object feature vectors are statistically modeled with normal distribution and the background feature vectors with uniform distribution, which results in a statistical object model. Although these steps, denoted in Figure 4.1 with ▲, are discussed in detail in Section 4.2, their short revision for the case of color modeling follows in Section 4.3.

In Section 4.3 the statistical object modeling in the case of color images is explained. First, the original training images are converted into color images, in which six-dimensional local feature vectors are computed. Figure 4.1 shows these steps by boxes marked with ♦. The definition of the object area and the statistical modeling, which are presented in detail in Section 4.2, are here reviewed for the case of six-dimensional color feature vectors. Figure 4.1 denotes this part of the training phase by boxes marked with ▲.

In the recognition phase, which is described in Chapter 5, the system is able to deal with multi-object scenes. If there is information about the environment, in which a multi-object scene is recorded, it is helpful to use this additional knowledge. Section 4.4 presents how this context

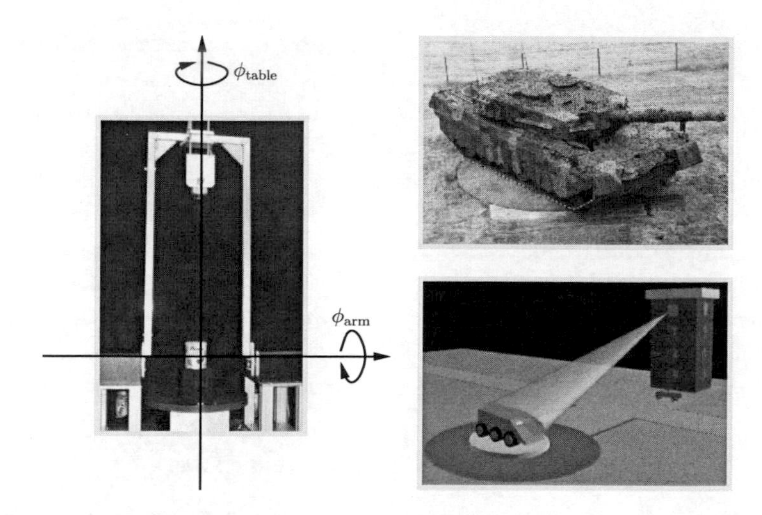

Figure 4.2: Left: turntable for laboratory applications. Right: turntable for big objects.

information can be learned in the training phase. This part of the system is marked in Figure 4.1 with the symbol ▼.

4.1 Training Data

Under training data both the training images $\boldsymbol{f}_{\kappa,\rho}$ and the poses $(\boldsymbol{\phi}_{\kappa,\rho}, \boldsymbol{t}_{\kappa\rho})$ of the objects in the training images are understood. The training data can be collected with a turntable (Section 4.1.1), or by a hand-held camera (Section 4.1.2). The index κ denoting the number of class is omitted in this section, because the training data collection is identical for all objects $\Omega = \{\Omega_1, \Omega_2, \ldots, \Omega_\kappa, \ldots, \Omega_{N_\Omega}\}$.

4.1.1 Training Data Collection with Turntable

Figure 4.2 presents a turntable for big objects on the right side, while on the left side a turntable for laboratory applications is shown. Objects are put on such a turntable, the turntable moves to known rotation angles ϕ_{table}, and training images $\boldsymbol{f}_\rho$ for all these angles are taken. This procedure delivers information about the external rotation $\phi_{y,\rho}$ defined in Section 1.1 (see Figure 1.2) for all object views. The known angles of the camera arm ϕ_{arm} yield the external rotation

$\phi_{x,\rho}$. The external translation (scaling) $t_{\text{ext},\rho} = t_{z,\rho}$ can be set with the zoom parameter of the camera, or by shifting the camera along the arm. In this way, objects can be taken from all top and sidewise views, whereas the external pose parameters $(\boldsymbol{\phi}_{\text{ext},\rho}, t_{\text{ext},\rho})$ of the objects are known for all training images $\boldsymbol{f}_\rho$. The translation of the objects in the image plane (internal translation) $\boldsymbol{t}_{\text{int},\rho} = (t_{x,\rho}, t_{y,\rho})^{\text{T}}$ as well as the internal rotation $\phi_{\text{int},\rho} = \phi_{z,\rho}$ can be determined after the acquisition process using pixel values in the images. The object pose parameters are usually given relative to each other, which can be seen in Figure 1.2 (Section 1.1). One image $\boldsymbol{f}_{\rho_1}$ for each object class is used as a reference image. By pose of the object in the image $\boldsymbol{f}_{\rho_2}$, the 3D transformation (rotation and translation) that maps the object in the reference image $\boldsymbol{f}_{\rho_1}$ to the object in $\boldsymbol{f}_{\rho_2}$ is denoted. Concluding, using a turntable with a camera arm in the training phase, both the training images $\boldsymbol{f}_\rho$ and the object pose parameters $(\boldsymbol{\phi}_\rho, \boldsymbol{t}_\rho)$ are collected.

4.1.2 Training Data Collection with Hand-Held Camera

In the laboratory environment, training images can be taken with a special setup like a turntable with a camera arm, which is described in the previous section. In real problems of object recognition, it is much easier to record the objects using a hand-held camera (see Figure 4.3, left). After the image acquisition in this way, an image sequence for each object class is given, but the object poses in the images are unknown. The context modeling does not require any pose parameters, but in order to create the object models, it is necessary to estimate the internal and external object pose parameters for all frames [Grz04b]. For this purpose, first, the so-called structure-from-motion algorithm [Hei04], which recovers the camera parameters and scene geometry, is applied. Second, the camera poses are converted into the object pose parameters, which are defined according to Figure 1.2 in Section 1.1. Under camera parameters, both *intrinsic* and *extrinsic* parameters are understood. The intrinsic parameters describe physical properties of the camera itself; the extrinsic camera parameters denote its pose, i. e., position and orientation.

Given an image sequence, the structure-from-motion algorithm starts with the point feature detection and tracking. The feature detection algorithms which are used for this are generally intensity based. They define an interest operator, which considers the intensity values within a rectangular window of the image. Examples of such operators can be found in [Mor77, För87, Har88]. A survey and evaluation of these and other interest operators are given in [Sch00]. In this work, point features defined by Tomasi and Kanade [Tom91] are applied. After the point features were detected in the first image, they are tracked through the whole image sequence or its subsequence. The basis of this tracking technique is the so-called Lucas-Kanade tracker [Luc81]. Being originally intended for image registration, it uses a gradient descent minimization to find the best match for each feature independently. The error, which has to be minimized, is the difference between the intensity values in a window around the feature in the current and in the next image. The tracked point features are stored in a measurement matrix $\boldsymbol{\Gamma}$ one above another. Given N_q point features $\boldsymbol{q}_{i,j}$ observed in N_F frames, the measurement matrix $\boldsymbol{\Gamma}$ can be written as

$$\boldsymbol{\Gamma} = \begin{pmatrix} \boldsymbol{q}_{1,1} & \boldsymbol{q}_{1,2} & \cdots & \boldsymbol{q}_{1,N_q} \\ \boldsymbol{q}_{2,1} & \boldsymbol{q}_{2,2} & \cdots & \boldsymbol{q}_{2,N_q} \\ \vdots & \vdots & & \vdots \\ \boldsymbol{q}_{N_F,1} & \boldsymbol{q}_{N_F,2} & \cdots & \boldsymbol{q}_{N_F,N_q} \end{pmatrix} \quad . \tag{4.1}$$

Note that $\boldsymbol{\Gamma} \in \mathbb{R}^{2N_F \times N_q}$, because the point features $\boldsymbol{q}_{i,j}$ are two-dimensional vectors. Usually the number of frames N_F, in which all the N_q point features are visible and can be tracked, is less than the number of all training images N_ρ. Thus, the structure-from-motion algorithm is first performed for a subsequence of $N_F < N_\rho$ images, and then completed for the remaining $N_\rho - N_F$ images.

The measurement matrix $\boldsymbol{\Gamma}$ containing all point correspondences in the image sequence can be decomposed into the 3D structure of the scene $\boldsymbol{S}$ and the camera motion $\boldsymbol{M}$

$$\boldsymbol{\Gamma} = \boldsymbol{M}\boldsymbol{S} \quad . \tag{4.2}$$

Matrix $\boldsymbol{S} \in \mathbb{R}^{3 \times N_q}$ contains 3D coordinates of all N_q point features, matrix $\boldsymbol{M} \in \mathbb{R}^{2N_F \times 3}$ comprises the camera pose parameters for all N_F frames. This decomposition is called *factorization* and was first introduced by Tomasi and Kanade [Tom92]. Such a decomposition is generally not unique. For any invertible matrix $\boldsymbol{A} \in \mathbb{R}^{3 \times 3}$, the measurement matrix $\boldsymbol{\Gamma}$ can be expressed as

$$\boldsymbol{\Gamma} = (\boldsymbol{M}\boldsymbol{A})(\boldsymbol{A}^{-1}\boldsymbol{S}) \quad . \tag{4.3}$$

However, under certain assumptions a unique determination of the structure $\boldsymbol{S}$ and motion $\boldsymbol{M}$ for a particular image sequence is possible [Hei04]. An extensive summary of factorization methods can be found in [Kan98].

A common restriction to the factorization methods is that they are only able to obtain the camera motion $\boldsymbol{M}$, but not its intrinsic parameters like, e. g., its focal length. The techniques applied for this purpose are called *self-calibration* or *auto-calibration*. The first solution to this problem, which uses the Kruppa equations, was introduced by Faugeras [Fau92] and refined in [Zel96]. Another approach which makes use of the fundamental matrix of two views is described in [Men99] and evaluated in [Fus01]. Within the scope of this work, the approach of Hartley

[Har93] is introduced. It is based on a minimization of the back-projection error of reconstructed 3D points and camera parameters by Levenberg-Marquardt optimization. A detailed description of the self-calibration algorithms can be found in [Pol99, Fus00, Har03].

On the one hand, for the training of an object model not only a few images of the object, but often several hundred images from totally different viewpoints are needed. On the other hand, for the factorization methods described above (4.2), all detected point features have to be visible in all frames. Therefore, the factorization is commonly started for an initial subsequence of N_F images, through which all features can be tracked. The subsequence is chosen to be the longest one with a certain number of features visible in all its images. Then the system estimates the remaining camera pose parameters by non-linearly minimizing of the back-projection error image by image. For this, previously unused feature correspondences are triangulated after each estimation of a new camera pose to increase the number of available 3D points. As an initialization for each new image the projection matrix of a neighboring image in the sequence is used. The method is explained in detail in [Hei04]. At this point, the camera parameters for each image are given as projection matrices

$$\boldsymbol{P} = \boldsymbol{K}(\boldsymbol{R}^{\mathrm{T}}| - \boldsymbol{R}^{\mathrm{T}}\boldsymbol{t}_c) \quad , \tag{4.4}$$

where $\boldsymbol{P} \in \mathbb{R}^{3\times4}$ is the projection matrix, $\boldsymbol{K} \in \mathbb{R}^{3\times3}$ contains the intrinsic camera parameters $\boldsymbol{R} \in \mathbb{R}^{3\times3}$ is the camera rotation matrix, and $\boldsymbol{t}_c \in \mathbb{R}^{3\times1}$ denotes the camera translation vector. The symbol $|$ means that the matrices on both of its sides are put together so that $\boldsymbol{R}^{\mathrm{T}}| - \boldsymbol{R}^{\mathrm{T}}\boldsymbol{t}_c \in \mathbb{R}^{3\times4}$.

From the camera rotation matrix $\boldsymbol{R}$ and the camera translation vector $\boldsymbol{t}_c$, the object pose parameters $(\boldsymbol{\phi}_\rho, \boldsymbol{t}_\rho)$ in terms of the system for object classification and localization (see Figure 1.2) can be determined. First, the origin of the coordinate system is translated into the center of mass of the object [Grz05c]. Since the object is placed on a black background in training images, the feature tracking algorithm is only able to track features on the object itself. Therefore, all 3D points contained in the structure matrix $\boldsymbol{S}$ are assumed to belong to the object. Thus, the centroid of the reconstructed 3D points is used as an approximation of the center of mass of the object. The external rotation parameters of the object $\boldsymbol{\phi}_{\mathrm{ext},\rho} = (\phi_{x,\rho}, \phi_{y,\rho})^{\mathrm{T}}$ can be easily computed using the rectangular triangles depicted on the right side of Figure 4.3. The first external rotation angle $\phi_{x,\rho}$ can be expressed with

$$\phi_{x,\rho} = \arcsin \frac{t_{c,y'}}{\sqrt{t_{c,x'}^2 + t_{c,y'}^2 + t_{c,z'}^2}} \quad , \tag{4.5}$$

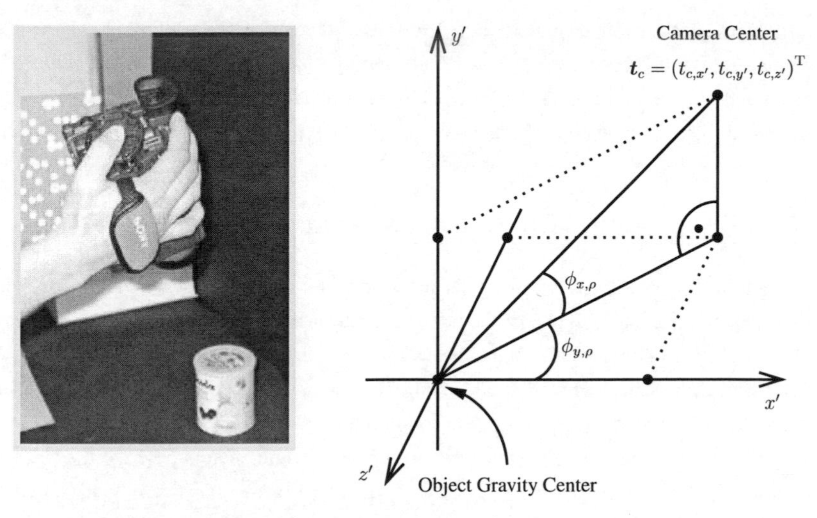

Figure 4.3: Left: image acquisition with hand-held camera. Right: conversion of camera pose parameters into object pose parameters, defined in Figure 1.2 in Section 1.1.

and the second external rotation angle $\phi_{y,\rho}$ with

$$\phi_{y,\rho} = \arcsin \frac{t_{c,z'}}{\sqrt{t^2_{c,x'} + t^2_{c,z'}}} \quad . \tag{4.6}$$

The external translation $t_{\text{ext},\rho} = t_{z,\rho}$ of the object can be determined from the distance of the camera to the object, which is given by

$$t_{z,\rho} = \sqrt{t^2_{c,x'} + t^2_{c,y'} + t^2_{c,z'}} \quad . \tag{4.7}$$

There are two possibilities to determine the internal object pose parameters in the images. First, the translation of the object in the image plane (internal translation) $\boldsymbol{t}_{\text{int},\rho} = (t_{x,\rho}, t_{y,\rho})^{\text{T}}$ as well as the internal rotation $\phi_{\text{int},\rho} = \phi_{z,\rho}$ can be estimated using pixel values in the images like in the case of image acquisition using a turntable (Section 4.1.1). Second, the information about the internal translation $\boldsymbol{t}_{\text{int},\rho} = (t_{x,\rho}, t_{y,\rho})^{\text{T}}$ of the object can be received by back-projection of the object gravity center into the image plane [Grz05c], while the internal object rotation $\phi_{\text{int},\rho} = \phi_{z,\rho}$ results from the projection of the y'-axis into the image plane (see Figure 4.3, right).

In this way, all data which are required for the next steps of the training phase (see Figure 4.1), namely the object images $\boldsymbol{f}_\rho$ and corresponding object poses $(\boldsymbol{\phi}_\rho, \boldsymbol{t}_\rho)$, are collected and ready for further processing. As mentioned at the end of Section 4.1.1, the object poses are represented relative to the object pose parameters in the reference image.

4.2 Object Model for Gray Level Images

This section discusses the steps of object model creation using gray level images. After the training data collection with one of the methods presented in Section 4.1, the system converts and resizes the original training RGB images into gray level images of size $2^n \times 2^n$ ($n \in \mathbb{N}$) pixels[2] with algorithms described in Section 3.1.3. The reason for the image size $2^n \times 2^n$ pixels becomes clear in Section 4.2.1, where the computation of feature vectors is explained. In Section 4.2.2, the decision is made which of the computed feature vectors describe the object, and which belong to the background. The object feature vectors are statistically modeled with the normal distribution (2.20), which is presented in Section 4.2.3. The statistical modeling of the background features with the uniform distribution (2.22) follows in Section 4.2.4. All components of the resulting statistical object model for gray level images are summarized in Section 4.2.5. Due to the fact that the object modeling using gray level images is identical for all objects $\Omega = \{\Omega_1, \Omega_2, \ldots, \Omega_\kappa, \ldots, \Omega_{N_\Omega}\}$ the index κ denoting the number of the object class and model is omitted in the present section, with the exception of the summary in Section 4.2.5.

4.2.1 Feature Extraction

For the object modeling using gray level images, objects are described by two-dimensional local feature vectors. As mentioned in Section 1.3, the main advantage of the local feature vectors is that a local disturbance affects only the features in a small region around it. In contrast to this, a global feature vector can totally change, if only one pixel in the image varies. Moreover, using the local feature vectors it is easier to distinguish the object part from the background part in the image, which becomes clear in Section 4.2.2. Finally, the algorithm for object classification and localization in multi-object scenes [Grz04a], which follows in Section 5.4, assigns the feature vectors to the objects found in the scene. This is not possible in the case of a global feature vector describing the whole image.

The system determines a set C_ρ of local feature vectors $\boldsymbol{c}_{\widehat{m},\rho}$ (3.14) for all N_ρ preprocessed

[2] Although the system is able to work with any value of $n \in \mathbb{N}$, usually $n \in \{6, 7, 8, 9\}$ is used.

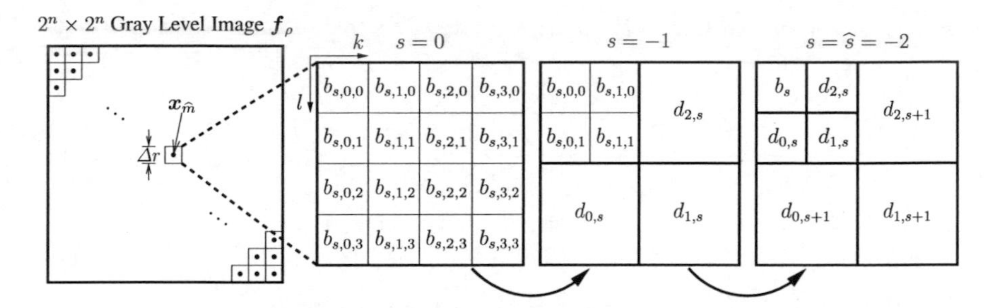

Figure 4.4: 2D signal decomposition with the wavelet transformation for a local neighborhood of size 4×4 pixels. The final coefficients result from gray values $b_{0,k,l}$ and have the following meaning: b_{-2} : low-pass horizontal and low-pass vertical, $d_{0,-2}$: low-pass horizontal and high-pass vertical, $d_{1,-2}$: high-pass horizontal and high-pass vertical, $d_{2,-2}$: high-pass horizontal and low-pass vertical.

training images $\boldsymbol{f}_{\rho=1,\ldots,N_\rho}$ of an object Ω using the discrete wavelet transformation defined in Section 2.2.3. For the calculation of $\boldsymbol{c}_{\widehat{m},\rho}$, a grid with the size $\Delta r = 2^{|\widehat{s}|}$, where $\widehat{s}$ is the minimum multiresolution scale parameter[3] s introduced in Section 2.2.2, is laid on the image $\boldsymbol{f}_\rho$ [Grz05b] (see Figure 4.4). A quadratic neighborhood of size $2^{|\widehat{s}|} \times 2^{|\widehat{s}|}$ pixels of the grid point $\boldsymbol{x}_{\widehat{m}} \in \mathbb{R}^2$ in the image $\boldsymbol{f}_\rho$ is treated as a two-dimensional discrete signal b_0, whereas $b_{0,k,l} \in \{0, 1, \ldots, 255\}$, and decomposed to low-pass and high-pass coefficients. One step of this decomposition was shown in Figure 2.7 and described in detail in Section 2.2.3. All grid points $\boldsymbol{x}_{\widehat{m}}$ can be written together as a set

$$X = \{\boldsymbol{x}_1, \ldots, \boldsymbol{x}_{\widehat{m}}, \ldots, \boldsymbol{x}_{\widehat{M}}\} \quad , \tag{4.8}$$

where

$$\widehat{M} = \frac{2^n \cdot 2^n}{2^{|\widehat{s}|} \cdot 2^{|\widehat{s}|}} \quad . \tag{4.9}$$

The set C_ρ of local feature vectors $\boldsymbol{c}_{\widehat{m},\rho}$ for the training image $\boldsymbol{f}_\rho$ can be respectively denoted as

$$C_\rho = \{\boldsymbol{c}_{1,\rho}, \ldots, \boldsymbol{c}_{\widehat{m},\rho}, \ldots, \boldsymbol{c}_{\widehat{M},\rho}\} \quad . \tag{4.10}$$

First, the feature extraction for the resolution level $L_{\widehat{s}} = L_{-2}$, for which $\Delta r = 2^{|\widehat{s}|} = 4$, is discussed. Figure 4.4 illustrates this case. After the first step of signal decomposition, i. e., for

[3] Further decomposition of the signal with the wavelet transformation is not possible.

$s = -1$, the low-pass coefficients $b_{-1,k,l}$ can be expressed according to the equation (2.58) with

$$b_{-1,k,l} = \sum_{j=-\left\lfloor\frac{N_\psi-1}{2}\right\rfloor}^{\left\lceil\frac{N_\psi-1}{2}\right\rceil} h_j \sum_{i=-\left\lfloor\frac{N_\psi-1}{2}\right\rfloor}^{\left\lceil\frac{N_\psi-1}{2}\right\rceil} h_i b_{0,2k+i,2l+j} \quad , \tag{4.11}$$

where N_ψ is the number of wavelet coefficients g_k and corresponding scaling function coefficients h_k. The second step of the decomposition results in a low-pass horizontal and low-pass vertical (2.58)

$$b_{-2} = \sum_{j=-\left\lfloor\frac{N_\psi-1}{2}\right\rfloor}^{\left\lceil\frac{N_\psi-1}{2}\right\rceil} h_j \sum_{i=-\left\lfloor\frac{N_\psi-1}{2}\right\rfloor}^{\left\lceil\frac{N_\psi-1}{2}\right\rceil} h_i b_{-1,i,j} \quad , \tag{4.12}$$

low-pass horizontal and high-pass vertical (2.59)

$$d_{0,-2} = \sum_{j=-\left\lfloor\frac{N_\psi-1}{2}\right\rfloor}^{\left\lceil\frac{N_\psi-1}{2}\right\rceil} g_j \sum_{i=-\left\lfloor\frac{N_\psi-1}{2}\right\rfloor}^{\left\lceil\frac{N_\psi-1}{2}\right\rceil} h_i b_{-1,i,j} \quad , \tag{4.13}$$

high-pass horizontal and high-pass vertical (2.60)

$$d_{1,-2} = \sum_{j=-\left\lfloor\frac{N_\psi-1}{2}\right\rfloor}^{\left\lceil\frac{N_\psi-1}{2}\right\rceil} g_j \sum_{i=-\left\lfloor\frac{N_\psi-1}{2}\right\rfloor}^{\left\lceil\frac{N_\psi-1}{2}\right\rceil} g_i b_{-1,i,j} \quad , \tag{4.14}$$

and high-pass horizontal and low-pass vertical (2.61)

$$d_{2,-2} = \sum_{j=-\left\lfloor\frac{N_\psi-1}{2}\right\rfloor}^{\left\lceil\frac{N_\psi-1}{2}\right\rceil} h_j \sum_{i=-\left\lfloor\frac{N_\psi-1}{2}\right\rfloor}^{\left\lceil\frac{N_\psi-1}{2}\right\rceil} g_i b_{-1,i,j} \quad . \tag{4.15}$$

For this resolution level, namely $L_{\widehat{s}} = L_{-2}$, the two-dimensional local feature vector $\boldsymbol{c}_{\widehat{m},\rho}$ describing the neighborhood of the grid point $\boldsymbol{x}_{\widehat{m}}$ in the image $\boldsymbol{f}_\rho$ is defined with

$$\boldsymbol{c}_{\widehat{m},\rho} = \begin{pmatrix} c_{\widehat{m},\rho,1} \\ c_{\widehat{m},\rho,2} \end{pmatrix} = \begin{pmatrix} \ln(2^{-2}|b_{-2,\widehat{m},\rho}|) \\ \ln[2^{-2}(|d_{0,-2,\widehat{m},\rho}| + |d_{1,-2,\widehat{m},\rho}| + |d_{2,-2,\widehat{m},\rho}|)] \end{pmatrix} \quad . \tag{4.16}$$

Note that the results of the signal decomposition $b_{-2,\widehat{m},\rho}$, $d_{0,-2,\widehat{m},\rho}$, $d_{1,-2,\widehat{m},\rho}$, and $d_{2,-2,\widehat{m},\rho}$ are denoted here with additional indexes $\widehat{m}$ and ρ in comparison to Figure 4.4, because they correspond to the feature vector $\boldsymbol{c}_{\widehat{m},\rho}$ in the training image $\boldsymbol{f}_\rho$.

Generally, the system is able to compute local feature vectors for any resolution level $L_{\widehat{s}}$, whereas in practice (Chapter 6) $\widehat{s} \in \{-1, -2, -3\}$ is preferred [Grz05b]. For the computation of the local feature vector $\boldsymbol{c}_{\widehat{m},\rho}$ for the resolution level $L_{\widehat{s}}$ in the image $\boldsymbol{f}_\rho$, a neighborhood of size $2^{|\widehat{s}|} \times 2^{|\widehat{s}|}$ pixels of the grid point $\boldsymbol{x}_{\widehat{m}}$ is taken into consideration. First, the 2D signal decomposition with the discrete wavelet transformation is $|\widehat{s}|$ times performed for this neighborhood, which is explained in detail in Section 2.2.3. The resulting coefficients $b_{\widehat{s},\widehat{m},\rho}$ (2.58), $d_{0,\widehat{s},\widehat{m},\rho}$ (2.59), $d_{1,\widehat{s},\widehat{m},\rho}$ (2.60), and $d_{2,\widehat{s},\widehat{m},\rho}$ (2.61) are used for feature vector computation as follows

$$\boldsymbol{c}_{\widehat{m},\rho} = \begin{pmatrix} c_{\widehat{m},\rho,1} \\ c_{\widehat{m},\rho,2} \end{pmatrix} = \begin{pmatrix} \ln(2^{\widehat{s}}|b_{\widehat{s},\widehat{m},\rho}|) \\ \ln[2^{\widehat{s}}(|d_{0,\widehat{s},\widehat{m},\rho}| + |d_{1,\widehat{s},\widehat{m},\rho}| + |d_{2,\widehat{s},\widehat{m},\rho}|)] \end{pmatrix} \quad . \tag{4.17}$$

The first component $c_{\widehat{m},\rho,1}$ of the feature vector $\boldsymbol{c}_{\widehat{m},\rho}$ in the image $\boldsymbol{f}_\rho$ depends on the low-pass coefficient $b_{\widehat{s},\widehat{m},\rho}$ and contains information about the pixel gray level values in the local neighborhood of the grid point $\boldsymbol{x}_{\widehat{m}}$. In the second element $c_{\widehat{m},\rho,2}$, information about discontinuities, e. g., edges or brightness changes, is stored. There is an additional assumption made for the system stability. If $c_{\widehat{m},\rho,1} < \ln(0.5)$ or $c_{\widehat{m},\rho,2} < \ln(0.5)$, they are set to $\ln(0.5)$. In the case of real digital images such a situation occurs very seldom. The mathematical operations used in the equations (4.16) and (4.17) improve the robustness of the feature vectors [Rei04]. The natural logarithm decreases the sensibility of the system to illumination changes and muffles any noises, which occur very often, especially in the real world environment. The multiplication of the coefficients by $2^{\widehat{s}}$ reduces the system dependency on the number $\widehat{s}$ of performed signal decomposition steps, and consequently on the resolution level $L_{\widehat{s}}$.

In order to apply the equations (4.16) and (4.17) for the feature vector computation, the wavelet (high-pass filter) ψ with the coefficients $g_{k=1,\ldots,N_\psi}$ and the scaling function (low-pass filter) φ with the coefficients $h_{k=1,\ldots,N_\psi}$ must be determined. In practice, the knowledge of one of these functions is sufficient, because coefficients of the other one can be computed applying the formula (2.55)

$$h_k = (-1)^k g_{1-k} \quad . \tag{4.18}$$

Pösl proved in [Pös99] that the best classification and localization rates are obtained with the *Johnston* wavelet ψ and the corresponding scaling function φ, which coefficients are listed in Table 4.1. Therefore, the Johnston wavelet and its corresponding scaling function are used for the object feature extraction in the scope of the present work.

k	g_k	h_k
-3	0.015274	0.015274
-2	0.099917	-0.099917
-1	0.098186	0.098186
0	-0.692937	0.692937
1	0.692937	0.692937
2	-0.098186	0.098186
3	-0.099917	-0.099917
4	-0.015274	0.015274

Table 4.1: Coefficients g_k of the Johnston wavelet ψ and coefficients h_k of the Johnston scaling function φ.

The feature vectors computed on the grid point $\boldsymbol{x}_{\widehat{m}}$ in all training images $\boldsymbol{f}_{\rho=1,\ldots,N_\rho}$ of an object Ω can be written together as a series

$$\boldsymbol{c}_{\widehat{m}} = \left(\boldsymbol{c}_{\widehat{m},1}, \boldsymbol{c}_{\widehat{m},2}, \ldots, \boldsymbol{c}_{\widehat{m},\rho}, \ldots, \boldsymbol{c}_{\widehat{m},N_\rho}\right) \quad . \tag{4.19}$$

The grid point $\boldsymbol{x}_{\widehat{m}}$ is located on the same place for all N_ρ training images $\boldsymbol{f}_\rho$.

4.2.2 Object Area

As one can imagine, some feature vectors in each training image describe the object, others belong to the background. In the real world environment, it cannot be assumed that the background in the recognition phase is a-priori known. Therefore, for the statistical object modeling, only feature vectors describing the object should be considered. Since the object takes usually only a part of the image, a tightly enclosing bounding region $O \subset X$ in the following called *object area* for each object class Ω is defined. Subsequently, the local feature vectors inside this object area are counted to the object and called *object feature vectors*, while the features outside O are denoted as *background feature vectors*. Under C_O the set of object feature vectors is understood. The object area can change its location, orientation, and size from image to image, which is shown in Figure 4.5.

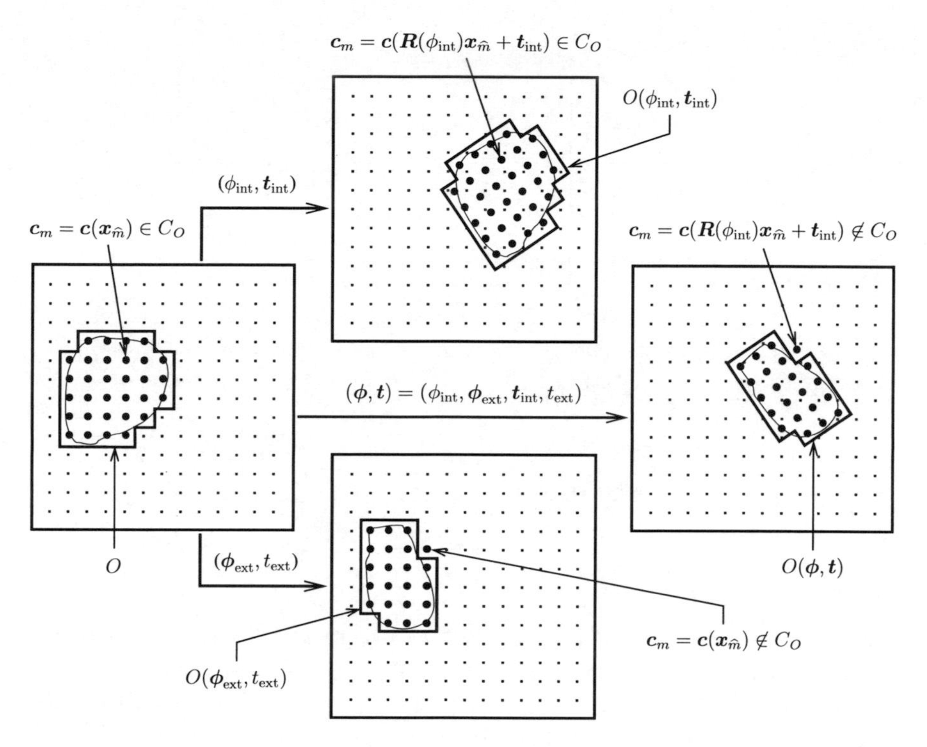

Figure 4.5: Left: object area in a reference training image. Top: object area after transformation with internal parameters. Bottom: object area after transformation with external parameters. Right: object area after transformation with internal and external parameters.

Internal Pose Parameters $(\phi_{\text{int}}, t_{\text{int}})$

In the simple case, when the object is rotated by $\phi_{\text{int}} \in \mathbb{R}$ around the perpendicular axis to the image plane and translated by $t_{\text{int}} \in \mathbb{R}^2$ in the image plane, its appearance and size do not change. However, the grid inside the object changes its location and orientation (see Figure 4.5). The new grid point positions x_m are calculated by

$$x_m = R(\phi_{\text{int}})x_{\hat{m}} + t_{\text{int}} \quad , \tag{4.20}$$

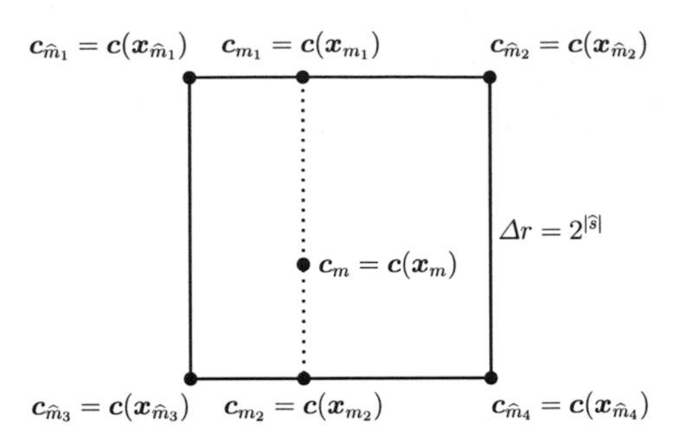

Figure 4.6: Bilinear interpolation of the object feature vector $\boldsymbol{c}_m$ from the four neighboring image feature vectors $\boldsymbol{c}_{\widehat{m}_1}$, $\boldsymbol{c}_{\widehat{m}_2}$, $\boldsymbol{c}_{\widehat{m}_3}$, and $\boldsymbol{c}_{\widehat{m}_4}$.

where $\boldsymbol{R}(\phi_{\mathrm{int}}) \in \mathbb{R}^{2\times 2}$ is the rotation matrix

$$\boldsymbol{R}(\phi_{\mathrm{int}}) = \begin{pmatrix} \cos(\phi_{\mathrm{int}}) & \sin(\phi_{\mathrm{int}}) \\ -\sin(\phi_{\mathrm{int}}) & \cos(\phi_{\mathrm{int}}) \end{pmatrix} \quad . \tag{4.21}$$

If the positions $\boldsymbol{x}_m$ of the object grid do not coincide with the positions $\boldsymbol{x}_{\widehat{m}}$ of the image grid, the object feature vector $\boldsymbol{c}_m$ on the transformed positions $\boldsymbol{x}_m$ is bilinear interpolated from the four adjacent image feature vectors $\boldsymbol{c}_{\widehat{m}_1}$, $\boldsymbol{c}_{\widehat{m}_2}$, $\boldsymbol{c}_{\widehat{m}_3}$, and $\boldsymbol{c}_{\widehat{m}_4}$ shown in Figure 4.6. First, the linear interpolation in horizonal direction is performed and delivers the following results

$$\boldsymbol{c}_{m_1} = \frac{||\boldsymbol{x}_{\widehat{m}_2} - \boldsymbol{x}_{m_1}||}{||\boldsymbol{x}_{\widehat{m}_2} - \boldsymbol{x}_{\widehat{m}_1}||}\boldsymbol{c}_{\widehat{m}_1} + \frac{||\boldsymbol{x}_{m_1} - \boldsymbol{x}_{\widehat{m}_1}||}{||\boldsymbol{x}_{\widehat{m}_2} - \boldsymbol{x}_{\widehat{m}_1}||}\boldsymbol{c}_{\widehat{m}_2} \tag{4.22}$$

and

$$\boldsymbol{c}_{m_2} = \frac{||\boldsymbol{x}_{\widehat{m}_4} - \boldsymbol{x}_{m_2}||}{||\boldsymbol{x}_{\widehat{m}_4} - \boldsymbol{x}_{\widehat{m}_3}||}\boldsymbol{c}_{\widehat{m}_3} + \frac{||\boldsymbol{x}_{m_2} - \boldsymbol{x}_{\widehat{m}_3}||}{||\boldsymbol{x}_{\widehat{m}_4} - \boldsymbol{x}_{\widehat{m}_3}||}\boldsymbol{c}_{\widehat{m}_4} \quad , \tag{4.23}$$

where the distance $||\boldsymbol{x}_{\widehat{m}_4} - \boldsymbol{x}_{\widehat{m}_3}||$ between the grid points $\boldsymbol{x}_{\widehat{m}_4}$ and $\boldsymbol{x}_{\widehat{m}_3}$ is equal to Δr. The final result $\boldsymbol{c}_m$ is given by a linear interpolation in the vertical direction

$$\boldsymbol{c}_m = \frac{||\boldsymbol{x}_m - \boldsymbol{x}_{m_2}||}{||\boldsymbol{x}_{m_2} - \boldsymbol{x}_{m_1}||}\boldsymbol{c}_{m_1} + \frac{||\boldsymbol{x}_{m_1} - \boldsymbol{x}_m||}{||\boldsymbol{x}_{m_2} - \boldsymbol{x}_{m_1}||}\boldsymbol{c}_{m_2} \quad . \tag{4.24}$$

As can be seen in Figure 4.5, the object feature vector $\boldsymbol{c}_m$ and the image feature vector $\boldsymbol{c}_{\widehat{m}}$ (4.19) have different meanings. $\boldsymbol{c}_{\widehat{m}}$ is computed in all training images on the same image grid point $\boldsymbol{x}_{\widehat{m}}$, while $\boldsymbol{c}_m$ moves with the object in terms of internal pose parameters $(\phi_{\mathrm{int}}, \boldsymbol{t}_{\mathrm{int}})$ and is computed in all training images on the same object grid point $\boldsymbol{x}_m$. In the two-dimensional case $(\phi_{\text{int}}, \boldsymbol{t}_{\text{int}})$, the size of the object is modeled as fixed and the object area O can be trained using only one image of the object. In order to determine the set C_O and thereby also the object area O, it is decided for each feature vector $\boldsymbol{c}_m$ either it belongs to the object ($\boldsymbol{c}_m \in C_O$) or to the background ($\boldsymbol{c}_m \notin C_O$). This decision is made by constant assignment functions ξ_m defined for all feature vectors $\boldsymbol{c}_m$ with the following rule

$$\begin{cases} \xi_m = 1 & \Rightarrow \quad \boldsymbol{c}_m \in C_O \\ \xi_m = 0 & \Rightarrow \quad \boldsymbol{c}_m \notin C_O \end{cases} \quad . \tag{4.25}$$

The assignment functions ξ_m are defined with

$$\xi_m = \begin{cases} 1, & \text{if} \quad c_{m,1} \geq S_\xi \\ 0, & \text{if} \quad c_{m,1} < S_\xi \end{cases} \quad , \tag{4.26}$$

where S_ξ is a threshold value. In practice the objects are taken in the training phase on homogeneous black background. After the computation of a background feature vector $\boldsymbol{c}_m \notin C_O$ the low-pass coefficient $b_{\widehat{s},m}$ (2.58) is close to zero

$$b_{\widehat{s},m} \rightarrow 0^+ \quad \Rightarrow \quad 2^{\widehat{s}}|b_{\widehat{s},m}| \rightarrow 0^+ \quad . \tag{4.27}$$

Hence, the first component $c_{m,1}$ of the feature vector $\boldsymbol{c}_m$ takes a very low value

$$c_{m,1} = \ln(2^{\widehat{s}}|b_{\widehat{s},m}|) \rightarrow -\infty \quad . \tag{4.28}$$

In contrast to this, the first components of the object feature vectors $\boldsymbol{c}_m \in C_O$ are usually much greater. The optimal threshold value S_ξ is determined manually for each object in the training phase. Very dark objects are taken on white background, and the inequations in (4.26) change their directions. Having the set C_O (4.25), the set of grid points belonging to the object O (object area) is also known.

External Pose Parameters $(\phi_{\text{ext}}, t_{\text{ext}})$

If an object is transformed with the external pose parameters, not only its size, but also its appearance, i. e., pixel values in the object area, can change. For some external transformations $(\phi_{\text{ext}}, t_{\text{ext}})$ a local feature vector $\boldsymbol{c}_m$ describes the object ($\boldsymbol{c}_m \in C_O$), for others it belongs to the background ($\boldsymbol{c}_m \notin C_O$), which is illustrated in Figure 4.5. In this case, the functions ξ_m (4.26) assigning the feature vectors $\boldsymbol{c}_m$ either to the object area or to the background (4.25) cannot be assumed to be constant. They have to be modeled as functions of external pose parameters. First, for all training images $\boldsymbol{f}_\rho$ of an object Ω taken from different viewpoints $(\boldsymbol{\phi}_{\text{ext},\rho}, \boldsymbol{t}_{\text{ext},\rho})$ the set of object feature vectors is determined in the same way as in the two-dimensional case (4.25). After that, the assignment functions ξ_m for the feature vectors $\boldsymbol{c}_m$ can be represented as discrete functions

$$\xi_m = \xi_m(\boldsymbol{\phi}_{\text{ext},\rho}, \boldsymbol{t}_{\text{ext},\rho}) \tag{4.29}$$

defined for all training viewpoints $(\boldsymbol{\phi}_{\text{ext},\rho}, \boldsymbol{t}_{\text{ext},\rho})$. However, in the recognition phase (Chapter 5), objects occur not only in the training poses, but also between them. In order to localize such objects, assignment functions defined on a continuous external pose parameter domain $(\boldsymbol{\phi}_{\text{ext}}, \boldsymbol{t}_{\text{ext}})$ are required

$$\xi_m = \xi_m(\boldsymbol{\phi}_{\text{ext}}, \boldsymbol{t}_{\text{ext}}) \quad . \tag{4.30}$$

For this purpose, each discrete assignment function (4.29) is approximated by a continuous assignment function (4.30). The theory of the function approximation was discussed in detail in Section 2.3. According to (2.64), the continuous assignment function $\xi_m(\boldsymbol{\phi}_{\text{ext}}, \boldsymbol{t}_{\text{ext}})$ can be expressed as a weighted sum of N_ξ basis functions $v_i(\boldsymbol{\phi}_{\text{ext}}, \boldsymbol{t}_{\text{ext}})$ with the weights $a_{m,i}$

$$\xi_m(\boldsymbol{\phi}_{\text{ext}}, \boldsymbol{t}_{\text{ext}}) = \sum_{i=0}^{N_\xi-1} a_{m,i} v_i(\boldsymbol{\phi}_{\text{ext}}, \boldsymbol{t}_{\text{ext}}) \quad . \tag{4.31}$$

This is made using the sine-cosine (2.70) or cosine (2.71) transformation discussed in Section 2.2.1 and Section 2.3. The values of the discrete function $\xi_m(\boldsymbol{\phi}_{\text{ext},\rho}, \boldsymbol{t}_{\text{ext},\rho})$, which has to be approximated, are known for the finite number N_ρ of points, namely for all training views. Using this knowledge and (4.31), the following system of equations can be constructed

$$\begin{cases} \xi_m(\boldsymbol{\phi}_{\text{ext},1}, \boldsymbol{t}_{\text{ext},1}) - \sum\limits_{i=0}^{N_\xi-1} a_{m,i} v_i(\boldsymbol{\phi}_{\text{ext},1}, \boldsymbol{t}_{\text{ext},1}) = 0 \\ \xi_m(\boldsymbol{\phi}_{\text{ext},2}, \boldsymbol{t}_{\text{ext},2}) - \sum\limits_{i=0}^{N_\xi-1} a_{m,i} v_i(\boldsymbol{\phi}_{\text{ext},2}, \boldsymbol{t}_{\text{ext},2}) = 0 \\ \qquad \cdots \\ \xi_m(\boldsymbol{\phi}_{\text{ext},N_\rho}, \boldsymbol{t}_{\text{ext},N_\rho}) - \sum\limits_{i=0}^{N_\xi-1} a_{m,i} v_i(\boldsymbol{\phi}_{\text{ext},N_\rho}, \boldsymbol{t}_{\text{ext},N_\rho}) = 0 \end{cases} \quad . \tag{4.32}$$

If $N_\rho \geq N_\xi - 1$ and the equations are linearly independent, the weights $a_{m,i=0,\ldots,N_\xi-1}$ can be determined with this system of equations. In this way, the continuous functions $\xi_m = \xi_m(\boldsymbol{\phi}_{\text{ext}}, \boldsymbol{t}_{\text{ext}})$

for all feature vectors $\boldsymbol{c}_m$ are given. However, their value set is no more limited to $\{0, 1\}$. The functions ξ_m can take each value in the range $[0, 1]$, sometimes even a little more than one, or less than zero. Therefore, the decision rule (4.25) cannot be applied in this form. In this case, the decision is made for all external pose parameters $(\boldsymbol{\phi}_{\text{ext}}, \boldsymbol{t}_{\text{ext}})$ as follows

$$\begin{cases} \xi_m(\boldsymbol{\phi}_{\text{ext}}, \boldsymbol{t}_{\text{ext}}) \geq S_O \;\Rightarrow\; \boldsymbol{c}_m \in C_O \\ \xi_m(\boldsymbol{\phi}_{\text{ext}}, \boldsymbol{t}_{\text{ext}}) < S_O \;\Rightarrow\; \boldsymbol{c}_m \notin C_O \end{cases} , \tag{4.33}$$

where the threshold value S_O is set manually and has the same value for all object classes Ω.

Internal and External Pose Parameters $(\boldsymbol{\phi}, \boldsymbol{t})$

On the one hand, the object area O can be transformed with the internal pose parameters $(\phi_{\text{int}}, \boldsymbol{t}_{\text{int}})$ and considered as their continuous function

$$O = O(\phi_{\text{int}}, \boldsymbol{t}_{\text{int}}) \quad , \tag{4.34}$$

whereas the new grid point positions $\boldsymbol{x}_m$ are computed with (4.20). The set C_O of the object feature vectors $\boldsymbol{c}_m \in C_O$ is constant in this case. On the other hand, using (4.33), the object area O can be determined even for viewpoints $(\boldsymbol{\phi}_{\text{ext}}, \boldsymbol{t}_{\text{ext}})$, which are not taken into account in the training phase

$$O = O(\boldsymbol{\phi}_{\text{ext}}, \boldsymbol{t}_{\text{ext}}) \quad . \tag{4.35}$$

The set of the object feature vectors changes with the external pose parameters too

$$C_O = C_O(\boldsymbol{\phi}_{\text{ext}}, \boldsymbol{t}_{\text{ext}}) \quad . \tag{4.36}$$

Considering both internal and external transformation parameters, the object area O can be expressed as a function

$$O = O(\boldsymbol{\phi}, \boldsymbol{t}) \tag{4.37}$$

defined on a continuous pose parameter domain $(\boldsymbol{\phi}, \boldsymbol{t})$.

4.2.3 Statistical Modeling of Object Features

In order to handle illumination changes and low-frequency noise, the elements $c_{m,q=1,2}$ of the local feature vectors $\boldsymbol{c}_m$ are interpreted as random variables (Definition 2.1). Assuming that the object feature vectors $\boldsymbol{c}_m \in C_O$ are statistically independent on the feature vectors outside the object area O, the background feature vectors $\boldsymbol{c}_m \notin C_O$ can be disregarded in this section and modeled separately (Section 4.2.4). For the statistical modeling of the object feature vectors $\boldsymbol{c}_m \in C_O$ their elements $c_{m,q}$ are considered as normal random variables and represented by following density functions (2.19)

$$p(c_{m,q}|\mu_{m,q}, \sigma_{m,q}, \boldsymbol{\phi}, \boldsymbol{t}) = \frac{1}{\sigma_{m,q}\sqrt{2\pi}} \exp\left(\frac{(c_{m,q} - \mu_{m,q})^2}{-2\sigma^2_{m,q}}\right) \quad , \tag{4.38}$$

where $\mu_{m,q}$ is the mean value (Definition 2.4) and $\sigma_{m,q}$ the standard deviation (Definition 2.5) for the feature vector element $c_{m,q}$. Assuming the statistical independence of the elements $c_{m,1}$ and $c_{m,2}$, which can be done due to their different interpretations in terms of signal processing (Section 4.2.1), the density function for the object feature vector $\boldsymbol{c}_m \in C_O$ can be written according to (2.21) as

$$p(\boldsymbol{c}_m|\boldsymbol{\mu}_m, \boldsymbol{\sigma}_m, \boldsymbol{\phi}, \boldsymbol{t}) = \prod_{q=1}^{2} p(c_{m,q}|\mu_{m,q}, \sigma_{m,q}, \boldsymbol{\phi}, \boldsymbol{t}) \quad , \tag{4.39}$$

where $\boldsymbol{\mu}_m = (\mu_{m,1}, \mu_{m,2})^{\mathrm{T}}$ is the mean value vector and $\boldsymbol{\sigma}_m = (\sigma_{m,1}, \sigma_{m,2})^{\mathrm{T}}$ the standard deviation vector. The interpretation of the formula (4.39) is the following: $p(\boldsymbol{c}_m|\boldsymbol{\mu}_m, \boldsymbol{\sigma}_m, \boldsymbol{\phi}, \boldsymbol{t})$ is the probability density to observe the feature vector $\boldsymbol{c}_m$ on the object grid point $\boldsymbol{x}_m \in O(\boldsymbol{\phi}, \boldsymbol{t})$. Further, it is supposed that the feature vectors belonging to the object $\boldsymbol{c}_m \in C_O$ are statistically independent on each other. Under this assumption, an object can be described by the probability density p to observe the feature vectors $\boldsymbol{c}_m \in C_O$ on the grid points $\boldsymbol{x}_m \in O(\boldsymbol{\phi}, \boldsymbol{t})$ as follows

$$p(C_O|\boldsymbol{B}, \boldsymbol{\phi}, \boldsymbol{t}) = \prod_{\boldsymbol{x}_m \in O} p(\boldsymbol{c}_m|\boldsymbol{\mu}_m, \boldsymbol{\sigma}_m, \boldsymbol{\phi}, \boldsymbol{t}) \quad , \tag{4.40}$$

where $\boldsymbol{B}$ comprehends the mean value vectors $\boldsymbol{\mu}_m$ and the standard deviation vectors $\boldsymbol{\sigma}_m$. This probability density is called *object density* and, taking into account (4.39), can be written more detailed as

$$p(C_O|\boldsymbol{B}, \boldsymbol{\phi}, \boldsymbol{t}) = \prod_{\boldsymbol{x}_m \in O} \prod_{q=1}^{2} p(c_{m,q}|\mu_{m,q}, \sigma_{m,q}, \boldsymbol{\phi}, \boldsymbol{t}) \quad . \tag{4.41}$$

In reality, neighboring feature vectors might be statistically dependent, but considering the full neighborhood relationship, e. g., by a Markov Random Field [Rue05], leads to a very complex model. Besides, modeling a dependency between neighboring object feature vectors in a row [Pös99] gives worse results than the assumption of the statistical independence [Rei04]. Additionally, by the statistical independence nonuniform illumination changes can be handled very well, for example when the lighting direction varies and some object parts become brighter, while in the same time other parts get darker.

In order to complete the object description with the object density (4.41), the means $\mu_{m,q}$ and the standard deviations $\sigma_{m,q}$ for all object feature vectors $\boldsymbol{c}_m$ have to be learned. For this purpose N_ρ training images $\boldsymbol{f}_\rho$ of each object with the corresponding transformation parameters $(\boldsymbol{\phi}_\rho, \boldsymbol{t}_\rho)$ obtained in the data collection step (Section 4.1) are used. After the feature extraction (4.17) presented in Section 4.2.1, the object area $O(\boldsymbol{\phi}_\rho, \boldsymbol{t}_\rho)$ is determined according to (4.37) and, if required, the feature vectors $\boldsymbol{c}_m \in C_{O,\rho}$ inside this object area are interpolated (4.24) with the bilinear interpolation. The mean vectors $\boldsymbol{\mu}_m$, written concatenated as $\boldsymbol{\mu}$, and the standard deviation vectors $\boldsymbol{\sigma}_m$, concatenated written as $\boldsymbol{\sigma}$, can be estimated by the maximization of the object density (4.41) over all N_ρ training images [Pös99]

$$(\widehat{\boldsymbol{\mu}}, \widehat{\boldsymbol{\sigma}}) = \underset{(\boldsymbol{\mu},\boldsymbol{\sigma})}{\operatorname{argmax}} \prod_{\rho=1}^{N_\rho} p(C_{O,\rho}|\boldsymbol{B}, \boldsymbol{\phi}_\rho, \boldsymbol{t}_\rho) \quad . \tag{4.42}$$

Considering (4.41) the maximization term (4.42) can be rewritten as follows

$$(\widehat{\boldsymbol{\mu}}, \widehat{\boldsymbol{\sigma}}) = \underset{(\boldsymbol{\mu},\boldsymbol{\sigma})}{\operatorname{argmax}} \prod_{\rho=1}^{N_\rho} \prod_{\boldsymbol{x}_m \in O} \prod_{q=1}^{2} p(c_{m,\rho,q}|\mu_{m,q}, \sigma_{m,q}, \boldsymbol{\phi}_\rho, \boldsymbol{t}_\rho) \quad , \tag{4.43}$$

where $c_{m,\rho,q=1,2}$ are components of the feature vector $\boldsymbol{c}_m \in C_{O,\rho}$ extracted in the training image $\boldsymbol{f}_\rho$ inside the object area $O(\boldsymbol{\phi}_\rho, \boldsymbol{t}_\rho)$.

In the two-dimensional case, where the object pose is defined exclusively with the internal transformation parameters $(\phi_{\text{int}}, \boldsymbol{t}_{\text{int}})$, the maximization term (4.43) written for the mean value $\mu_{m,q}$ and the standard deviation $\sigma_{m,q}$ modifies to

$$(\widehat{\mu}_{m,q}, \widehat{\sigma}_{m,q}) = \underset{(\mu_{m,q},\sigma_{m,q})}{\operatorname{argmax}} \prod_{\rho=1}^{N_\rho} p(c_{m,\rho,q}|\mu_{m,q}, \sigma_{m,q}, \phi_{\text{int},\rho}, \boldsymbol{t}_{\text{int},\rho}) \quad , \tag{4.44}$$

which using (4.38) can be written as

$$(\widehat{\mu}_{m,q}, \widehat{\sigma}_{m,q}) = \underset{(\mu_{m,q},\sigma_{m,q})}{\operatorname{argmax}} \prod_{\rho=1}^{N_\rho} \left(\frac{1}{\sigma_{m,q}\sqrt{2\pi}} \exp\left(\frac{(c_{m,\rho,q} - \mu_{m,q})^2}{-2\sigma_{m,q}^2} \right) \right) \quad . \tag{4.45}$$

This leads to the well known estimation terms for the mean value

$$\widehat{\mu}_{m,q} = \frac{1}{N_\rho} \sum_{\rho=1}^{N_\rho} c_{m,\rho,q} \tag{4.46}$$

and the standard deviation

$$\widehat{\sigma}^2_{m,q} = \frac{1}{N_\rho} \sum_{\rho=1}^{N_\rho} (c_{m,\rho,q} - \widehat{\mu}_{m,q})^2 \quad . \tag{4.47}$$

Concluding, in the two-dimensional case the means $\mu_{m,q}$ and the standard deviations $\sigma_{m,q}$ are learned as constant, because a particular feature vector $\boldsymbol{c}_m$ describes always the same part of the object independent on the internal pose parameters $(\phi_{\text{int}}, \boldsymbol{t}_{\text{int}})$.

In the three-dimensional case, in which the objects are transformed with the internal $(\phi_{\text{int}}, \boldsymbol{t}_{\text{int}})$ and the external pose parameters $(\boldsymbol{\phi}_{\text{ext}}, t_{\text{ext}})$, it cannot be assumed that a grid point $\boldsymbol{x}_m$ belongs to the object area O for all N_ρ training images $\boldsymbol{f}_\rho$ of an object (see Figure 4.5). Therefore, the mean values $\mu_{m,q}$ and the standard deviations $\sigma_{m,q}$ are determined by the maximization of the object density (4.41) over all training images, for which the corresponding grid point $\boldsymbol{x}_m$ belongs to the object area O, i. e.,

$$\forall \rho : \boldsymbol{x}_m \in O(\boldsymbol{\phi}_\rho, \boldsymbol{t}_\rho) \quad . \tag{4.48}$$

The estimation terms (4.44) and (4.45) written for the full transformation parameter space $(\boldsymbol{\phi}, \boldsymbol{t})$ under condition (4.48) look like

$$(\widehat{\mu}_{m,q}, \widehat{\sigma}_{m,q}) = \underset{(\mu_{m,q}, \sigma_{m,q})}{\operatorname{argmax}} \prod_\rho p(c_{m,\rho,q} | \mu_{m,q}, \sigma_{m,q}, \boldsymbol{\phi}_\rho, \boldsymbol{t}_\rho) \tag{4.49}$$

and

$$(\widehat{\mu}_{m,q}, \widehat{\sigma}_{m,q}) = \underset{(\mu_{m,q}, \sigma_{m,q})}{\operatorname{argmax}} \prod_\rho \left(\frac{1}{\sigma_{m,q}\sqrt{2\pi}} \exp\left(\frac{(c_{m,\rho,q} - \mu_{m,q})^2}{-2\sigma^2_{m,q}} \right) \right) \quad . \tag{4.50}$$

Assuming that the standard deviations $\sigma_{m,q}$ are constant, i. e., do not depend on the transformation parameters $(\boldsymbol{\phi}, \boldsymbol{t})$, and taking into account the condition (4.48), the formula (4.50) can be modified to

$$(\widehat{\mu}_{m,q}, \widehat{\sigma}_{m,q}) = \underset{(\mu_{m,q}, \sigma_{m,q})}{\operatorname{argmax}} \left(\frac{1}{(\sigma_{m,q}\sqrt{2\pi})^{N_m}} \exp\left(\frac{1}{-2\sigma^2_{m,q}} \sum_\rho (c_{m,\rho,q} - \mu_{m,q})^2 \right) \right) \quad , \tag{4.51}$$

where N_m is the number of training images, for which the grid point $\boldsymbol{x}_m$ belongs to the object area O, whereas for 3D objects $N_m < N_\rho$. Due to the negative exponent in (4.51) and having regard to (4.48), the maximization term (4.51) can be replaced by a minimization as follows

$$\widehat{\mu}_{m,q} = \underset{\mu_{m,q}}{\operatorname{argmin}} \sum_{\rho} (c_{m,\rho,q} - \mu_{m,q})^2 \quad . \tag{4.52}$$

If the object is transformed with the external pose parameters $(\boldsymbol{\phi}_{\text{ext}}, \boldsymbol{t}_{\text{ext}})$, a particular feature vector $\boldsymbol{c}_m$ does not describe always the same part of the object (see Figure 4.5). For this reason, the mean values $\mu_{m,q=1,2}$ for this feature vector $\boldsymbol{c}_m$ are modeled as continuous functions defined on the external pose parameter domain $(\boldsymbol{\phi}_{\text{ext}}, \boldsymbol{t}_{\text{ext}})$. For this purpose, according to (2.64), the continuous mean value function $\mu_{m,q}(\boldsymbol{\phi}_{\text{ext}}, \boldsymbol{t}_{\text{ext}})$ is expressed as a weighted sum of N_μ basis functions $v_i(\boldsymbol{\phi}_{\text{ext}}, \boldsymbol{t}_{\text{ext}})$ with the weights $w_{m,q,i}$

$$\mu_{m,q}(\boldsymbol{\phi}_{\text{ext}}, \boldsymbol{t}_{\text{ext}}) = \sum_{i=0}^{N_\mu - 1} w_{m,q,i} v_i(\boldsymbol{\phi}_{\text{ext}}, \boldsymbol{t}_{\text{ext}}) \quad . \tag{4.53}$$

This is made using the sine-cosine (2.70) or cosine (2.71) transformation discussed in Section 2.2.1 and Section 2.3. Though for the internal pose parameters $(\boldsymbol{\phi}_{\text{int}}, \boldsymbol{t}_{\text{int}})$ the mean values $\mu_{m,q}$ are constant (4.46), they can be regarded as functions of all transformation parameters $(\boldsymbol{\phi}, \boldsymbol{t})$ and their approximations (4.53) modified to

$$\mu_{m,q}(\boldsymbol{\phi}, \boldsymbol{t}) = \sum_{i=0}^{N_\mu - 1} w_{m,q,i} v_i(\boldsymbol{\phi}, \boldsymbol{t}) \quad . \tag{4.54}$$

Considering (4.52), (4.54), and condition (4.48), the weights $w_{m,q,i}$, concatenated written as a vector $\boldsymbol{w}_{m,q}$, can be determined by minimizing of the following expression

$$\widehat{\boldsymbol{w}}_{m,q} = \underset{\boldsymbol{w}_{m,q}}{\operatorname{argmin}} \sum_{\rho} \left(c_{m,\rho,q} - \sum_{i=0}^{N_\mu - 1} w_{m,q,i} v_i(\boldsymbol{\phi}_\rho, \boldsymbol{t}_\rho) \right)^2 \quad . \tag{4.55}$$

As mentioned before, the standard deviations $\sigma_{m,q}$ are modeled as constant, and according to (4.47) can be computed with

$$\widehat{\sigma}^2_{m,q} = \frac{1}{N_\rho} \sum_{\rho} \left(c_{m,\rho,q} - \sum_{i=0}^{N_\mu - 1} w_{m,q,i} v_i(\boldsymbol{\phi}_\rho, \boldsymbol{t}_\rho) \right)^2 \quad . \tag{4.56}$$

4.2.4 Statistical Modeling of Background Features

As mentioned in Section 4.2.3, the background feature vectors $\boldsymbol{c}_m \notin C_O$ are assumed to be statistically independent on the feature vectors inside the object area O and can be modeled separately from $\boldsymbol{c}_m \in C_O$. The elements $c_{m,q=1,2}$ of the background feature vectors are also interpreted as random variables (Definition 2.1), but, in contrast to the object feature vector elements (Section 4.2.3), not as normally distributed. Since the background in the recognition phase (Chapter 5) is a-priori unknown, each possible value of the background feature vector element $c_{m,q}$ can be observed with the same probability Therefore, the background feature vector elements $c_{m,q}$ are modeled as uniform random variables (2.22) and their constant density functions

$$p(c_{m,q}) = \frac{1}{\max(c_{m,q}) - \min(c_{m,q})} \tag{4.57}$$

do not depend on the transformation parameters $(\boldsymbol{\phi}, \boldsymbol{t})$. Assuming the statistical independence of the elements $c_{m,1}$ and $c_{m,2}$, which can be done due to their different interpretations in terms of signal processing (Section 4.2.1), the density functions for the background feature vectors $\boldsymbol{c}_m \notin C_O$ can be written according to (2.21) as

$$p(\boldsymbol{c}_m) = \prod_{q=1}^{2} \frac{1}{\max(c_{m,q}) - \min(c_{m,q})} = p_b \quad , \tag{4.58}$$

where p_b is a constant value. Assuming that the gray level images are represented with pixels $f_{x,y} \in \{0, 1, \ldots, 255\}$ and the local feature vectors are computed with (4.17) using the Johnston wavelet, the background density p_b amounts approximately to $e^{-3.5}$. The motivation for the background modeling becomes clear in Chapter 5 describing the algorithm for object classification and localization in real world environment.

4.2.5 Summary

The object and the background modeling presented in Section 4.2 result in statistical object models $\mathcal{M} = \{\mathcal{M}_1, \mathcal{M}_2, \ldots, \mathcal{M}_\kappa, \ldots, \mathcal{M}_{N_\Omega}\}$ for gray level images, which are created for all object classes $\Omega = \{\Omega_1, \Omega_2, \ldots, \Omega_\kappa, \ldots, \Omega_{N_\Omega}\}$ considered in the recognition task. Each object model $\mathcal{M}_\kappa$ can be regarded as a continuous function defined on the transformation parameter domain $(\boldsymbol{\phi}, \boldsymbol{t})$

$$\mathcal{M}_\kappa = \mathcal{M}_\kappa(\boldsymbol{\phi}, \boldsymbol{t}) \quad . \tag{4.59}$$

The statistical object model $\mathcal{M}_\kappa$ of the object class Ω_κ determines for all object pose parameters $(\boldsymbol{\phi}, \boldsymbol{t})$ the set of grid points $\boldsymbol{x}_{\kappa,m}$ belonging to the object area

$$O_\kappa = O_\kappa(\boldsymbol{\phi}, \boldsymbol{t}) \quad . \tag{4.60}$$

The two-dimensional local feature vectors computed on these grid points $\boldsymbol{x}_{\kappa,m} \in O_\kappa(\boldsymbol{\phi}, \boldsymbol{t})$ are represented in the object model $\mathcal{M}_\kappa$ by normal density functions given with the mean vectors $\boldsymbol{\mu}_{\kappa,m}(\boldsymbol{\phi}, \boldsymbol{t})$ and the standard deviation vectors $\boldsymbol{\sigma}_{\kappa,m}$ as follows

$$\boldsymbol{\mu}_{\kappa,m} = \boldsymbol{\mu}_{\kappa,m}(\boldsymbol{\phi}, \boldsymbol{t}) = \begin{pmatrix} \mu_{\kappa,m,1}(\boldsymbol{\phi}, \boldsymbol{t}) \\ \mu_{\kappa,m,2}(\boldsymbol{\phi}, \boldsymbol{t}) \end{pmatrix} \quad , \quad \boldsymbol{\sigma}_{\kappa,m} = \begin{pmatrix} \sigma_{\kappa,m,1} \\ \sigma_{\kappa,m,2} \end{pmatrix} \quad . \tag{4.61}$$

Additionally, each object model $\mathcal{M}_\kappa$ stores the background density value p_b.

4.3 Object Model for Color Images

This section extends the object model for gray level images discussed in the previous section to the object model for color images. First, the original training images acquired in the data collection step (Section 4.1) are resized to RGB images of size $2^n \times 2^n$ ($n \in \mathbb{N}$) pixels[4]. Then six-dimensional local feature vectors are computed in the resized images with the wavelet transformation, which is explained in Section 4.3.1. Section 4.3.2 defines the set of feature vectors describing the object (object feature vectors). The remaining feature vectors belong to the background and are called background features. The statistical modeling of the object feature vectors with the normal distribution (2.20) follows in Section 4.3.3; while in Section 4.3.4, the statistical modeling of the background features using the uniform distribution (2.22) is presented. All components of the resulting statistical object model for color images are summarized in Section 4.3.5. As can be seen in Figure 4.1, the training using gray level images described in the previous section consists of the same steps as in the case of color modeling. Therefore, the present section is based on the results from Section 4.2. Due to the fact that the object modeling using color images is identical for all object classes $\Omega = \{\Omega_1, \Omega_2, \ldots, \Omega_\kappa, \ldots, \Omega_{N_\Omega}\}$, the index κ denoting the number of the object class and model is omitted in the present section, with the exception of the summary in Section 4.3.5.

[4]Although the system is able to work with any value of $n \in \mathbb{N}$, usually $n \in \{6, 7, 8, 9\}$ is used.

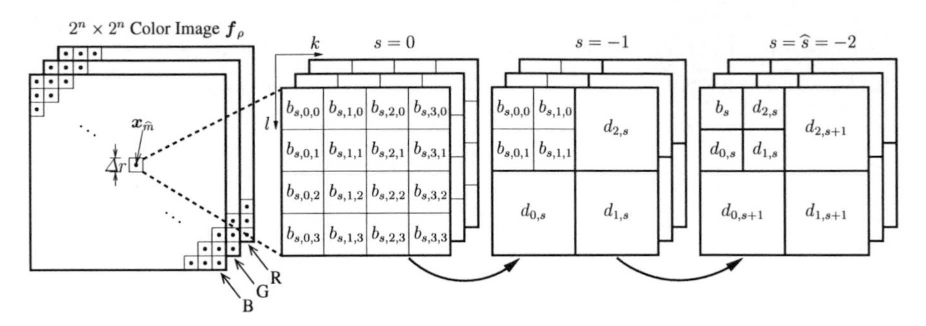

Figure 4.7: Wavelet decomposition for a local neighborhood of size 4×4 pixels done separately for the green, the red, and the blue channel. The final coefficients for the blue channel result from $b_{0,k,l}$ and have the following meaning: b_{-2} : low-pass horizontal and low-pass vertical, $d_{0,-2}$: low-pass horizontal and high-pass vertical, $d_{1,-2}$: high-pass horizontal and high-pass vertical, $d_{2,-2}$: high-pass horizontal and low-pass vertical.

4.3.1 Feature Extraction

In contrast to the object modeling using gray level images presented in Section 4.2, for the color modeling objects are described by six-dimensional local feature vectors. The feature extraction is performed separately for the red, the green, and the blue channel in the same way as for gray level images (Section 4.2.1). The first $c_{\widehat{m},\rho,1}$ and the second $c_{\widehat{m},\rho,2}$ element of the feature vector $\boldsymbol{c}_{\widehat{m},\rho}$ are calculated from the red channel, the third $c_{\widehat{m},\rho,3}$ and the fourth $c_{\widehat{m},\rho,4}$ from the green channel, and the fifth $c_{\widehat{m},\rho,5}$ and the sixth $c_{\widehat{m},\rho,6}$ from the blue channel [Grz05a]. Analogical to Section 4.2.1, the system lays a grid with the size $\Delta r = 2^{|\widehat{s}|}$ on the RGB image $\boldsymbol{f}_\rho$, which is for the resolution level $L_{\widehat{s}} = L_{-2}$ depicted in Figure 4.7. A quadratic neighborhood of size $2^{|\widehat{s}|} \times 2^{|\widehat{s}|}$ pixels of each grid point $\boldsymbol{x}_{\widehat{m}} \in \mathbb{R}^2$ is treated as a two-dimensional discrete signal for each channel separately and decomposed to low-pass and high-pass coefficients. This signal is denoted as $b_{r,0}$ for the red channel, $b_{g,0}$ for the green channel, and $b_{b,0}$ for the blue channel. Generally, the system is able to compute local feature vectors for any resolution level $L_{\widehat{s}}$, whereas in practice (Chapter 6) $\widehat{s} \in \{-1,-2,-3\}$ is preferred [Grz05b]. For this purpose, the 2D signal decomposition with the discrete wavelet transformation is $|\widehat{s}|$ times performed separately for $b_{r,0}$, $b_{g,0}$, and $b_{b,0}$ in the neighborhood of the grid point $\boldsymbol{x}_{\widehat{m}}$ (see Figure 4.7). One step of this decomposition is described in detail in Section 2.2.3. These two-dimensional signals are filtered with the Johnston wavelet ψ (high-pass filter) and the corresponding scaling function φ (low-pass filter), which values are listed in Table 4.1. The resulting coefficients $b_{\{r,g,b\},\widehat{s},\widehat{m},\rho}$ (2.58),

$d_{\{r,g,b\},0,\widehat{s},\widehat{m},\rho}$ (2.59), $d_{\{r,g,b\},1,\widehat{s},\widehat{m},\rho}$ (2.60), and $d_{\{r,g,b\},2,\widehat{s},\widehat{m},\rho}$ (2.61) are used for the feature vector computation as follows

$$\boldsymbol{c}_{\widehat{m},\rho} = \begin{pmatrix} c_{\widehat{m},\rho,1} \\ c_{\widehat{m},\rho,2} \\ c_{\widehat{m},\rho,3} \\ c_{\widehat{m},\rho,4} \\ c_{\widehat{m},\rho,5} \\ c_{\widehat{m},\rho,6} \end{pmatrix} = \begin{pmatrix} \ln(2^{\widehat{s}}|b_{r,\widehat{s},\widehat{m},\rho}|) \\ \ln[2^{\widehat{s}}(|d_{r,0,\widehat{s},\widehat{m},\rho}| + |d_{r,1,\widehat{s},\widehat{m},\rho}| + |d_{r,2,\widehat{s},\widehat{m},\rho}|)] \\ \ln(2^{\widehat{s}}|b_{g,\widehat{s},\widehat{m},\rho}|) \\ \ln[2^{\widehat{s}}(|d_{g,0,\widehat{s},\widehat{m},\rho}| + |d_{g,1,\widehat{s},\widehat{m},\rho}| + |d_{g,2,\widehat{s},\widehat{m},\rho}|)] \\ \ln(2^{\widehat{s}}|b_{b,\widehat{s},\widehat{m},\rho}|) \\ \ln[2^{\widehat{s}}(|d_{b,0,\widehat{s},\widehat{m},\rho}| + |d_{b,1,\widehat{s},\widehat{m},\rho}| + |d_{b,2,\widehat{s},\widehat{m},\rho}|)] \end{pmatrix} \quad . \tag{4.62}$$

The low-pass and high-pass coefficients resulting from the red channel are denoted with the index r, from the green channel with the index g, and from the blue channel with the index b. Similar to the feature computation for the gray level images (Section 4.2.1), the color feature vectors computed on the grid point $\boldsymbol{x}_{\widehat{m}}$ in all training images $\boldsymbol{f}_{\rho=1,\ldots,N_\rho}$ of an object Ω can be written together as a series

$$\boldsymbol{c}_{\widehat{m}} = (\boldsymbol{c}_{\widehat{m},1}, \boldsymbol{c}_{\widehat{m},2}, \ldots, \boldsymbol{c}_{\widehat{m},\rho}, \ldots, \boldsymbol{c}_{\widehat{m},N_\rho}) \quad . \tag{4.63}$$

The grid point $\boldsymbol{x}_{\widehat{m}}$ is located on the same place for all N_ρ training images $\boldsymbol{f}_\rho$.

4.3.2 Object Area

Some of the color feature vectors extracted with the wavelet signal decomposition in Section 4.3.1 describe the object, other belong to the background. The same as for the object modeling using gray level images presented in Section 4.2, in the case of the color modeling an object area $O \subset X$ is defined, the local feature vectors inside this object area are called object features, the features outside O are denoted as background feature vectors, and under C_O the set of object feature vectors is understood. The behavior of the object area O in the case of the internal $(\phi_{\text{int}}, \boldsymbol{t}_{\text{int}})$ and the external $(\boldsymbol{\phi}_{\text{ext}}, t_{\text{ext}})$ transformation parameters is described in detail in Section 4.2.2 and illustrated in Figure 4.5.

If the object is rotated by $\phi_{\text{int}} \in \mathbb{R}$ and translated by $\boldsymbol{t}_{\text{int}} \in \mathbb{R}^2$, its appearance and size do not change, but in some cases the object grid positions $\boldsymbol{x}_m$ are not coinciding with the positions $\boldsymbol{x}_{\widehat{m}}$ of the image grid (see Figure 4.5). The feature vectors $\boldsymbol{c}_m$ on these object grid points $\boldsymbol{x}_m$ are estimated using the bilinear interpolation (4.24) explained in Section 4.2.2 (see Figure 4.6). Due to determine the set C_O and thereby also the object area O, it is decided for each feature vector $\boldsymbol{c}_m$ either it belongs to the object ($\boldsymbol{c}_m \in C_O$) or to the background ($\boldsymbol{c}_m \notin C_O$). The same as for the object modeling using gray level images, this decision is made by constant assignment functions ξ_m defined for all feature vectors $\boldsymbol{c}_m$ with the rule (4.25), but the definition of the assignment functions ξ_m changes for the color modeling to

$$\xi_m = \begin{cases} 1, & \text{if } \frac{1}{3}\left(c_{m,1} + c_{m,3} + c_{m,5}\right) \geq S_\xi \\ 0, & \text{if } \frac{1}{3}\left(c_{m,1} + c_{m,3} + c_{m,5}\right) < S_\xi \end{cases} , \tag{4.64}$$

where S_ξ is a threshold value.

With the external transformation parameters $(\boldsymbol{\phi}_{\text{ext}}, \boldsymbol{t}_{\text{ext}})$, the size and the appearance of heterogeneous three-dimensional objects vary from image to image (see Figure 1.2). For some external transformations $(\boldsymbol{\phi}_{\text{ext}}, \boldsymbol{t}_{\text{ext}})$, a local feature vector $\boldsymbol{c}_m$ describes the object ($\boldsymbol{c}_m \in C_O$), for others it belongs to the background ($\boldsymbol{c}_m \notin C_O$) (see Figure 4.5). In this case, the assignment functions ξ_m (4.64) are not constant. They depend on the external pose parameters $(\boldsymbol{\phi}_{\text{ext}}, \boldsymbol{t}_{\text{ext}})$

$$\xi_m = \xi_m(\boldsymbol{\phi}_{\text{ext}}, \boldsymbol{t}_{\text{ext}}) \quad . \tag{4.65}$$

The determination of these functions by approximation of their discrete versions using the sine-cosine (2.70) or the cosine (2.71) transformation is explained in detail in Section 4.2.2. Their value set is no more limited to $\{0, 1\}$. The functions ξ_m can take each value in the range $[0, 1]$, sometimes even a little more than one, or less than zero. Therefore, the decision rule (4.25) cannot be applied in the three-dimensional case. Instead of that, the rule (4.33) is used.

The same as for the object modeling using gray level images, for the color modeling the object area O can be expressed as a function

$$O = O(\boldsymbol{\phi}, \boldsymbol{t}) \tag{4.66}$$

defined on a continuous pose parameter domain $(\boldsymbol{\phi}, \boldsymbol{t})$. It means that for all pose parameters $(\boldsymbol{\phi}, \boldsymbol{t})$, the set of grid points belonging to the object $\boldsymbol{x}_m \in O$ and thereby the set of feature vectors $\boldsymbol{c}_m \in C_O$ are known.

4.3.3 Statistical Modeling of Object Features

Analogical to the object modeling using gray level images (Section 4.2), the elements $c_{m,q=1,\ldots,6}$ of the color feature vectors $\boldsymbol{c}_m$ are also interpreted as random variables (Definition 2.1). The object feature vectors $\boldsymbol{c}_m \in C_O$ are assumed to be statistically independent of the background features $\boldsymbol{c}_m \notin C_O$ and modeled as normal density functions with (4.38). In reality, the elements $c_{m,q=1,\ldots,6}$ of the feature vectors $\boldsymbol{c}_m$ computed from the red, the green, and the blue channel correlate with each other, but modeling of this statistical dependency leads to a very complex model and does not yield better recognition rates. Therefore, within the scope of the present work, the statistical independence of $c_{m,q=1,\ldots,6}$ is assumed, and the density function for the object color feature vector $\boldsymbol{c}_m \in C_O$ can be expressed as follows

$$p(\boldsymbol{c}_m|\boldsymbol{\mu}_m, \boldsymbol{\sigma}_m, \boldsymbol{\phi}, \boldsymbol{t}) = \prod_{q=1}^{6} p(c_{m,q}|\mu_{m,q}, \sigma_{m,q}, \boldsymbol{\phi}, \boldsymbol{t}) \quad , \tag{4.67}$$

where $\boldsymbol{\mu}_m$ is the six-dimensional mean value vector and $\boldsymbol{\sigma}_m$ is the six-dimensional standard deviation vector. Under assumption of the statistical independence of the color feature vectors belonging to the object $\boldsymbol{c}_m \in C_O$, the object density (4.41) can be written as

$$p(C_O|\boldsymbol{B}, \boldsymbol{\phi}, \boldsymbol{t}) = \prod_{\boldsymbol{x}_m \in O} \prod_{q=1}^{6} p(c_{m,q}|\mu_{m,q}, \sigma_{m,q}, \boldsymbol{\phi}, \boldsymbol{t}) \quad , \tag{4.68}$$

where $\boldsymbol{B}$ comprehends the mean value vectors $\boldsymbol{\mu}_m$ and the standard deviation vectors $\boldsymbol{\sigma}_m$. In order to complete the object description with the object density (4.68) in the case of the color modeling the means $\mu_{m,q=1,\ldots,6}$ and the standard deviations $\sigma_{m,q=1,\ldots,6}$ are learned in the same way as presented in Section 4.2.3.

4.3.4 Statistical Modeling of Background Features

The background color feature vectors $\boldsymbol{c}_m \notin C_O$ are assumed to be statistically independent on the object features $\boldsymbol{c}_m \in C_O$, their elements $c_{m,q=1,\ldots,6}$ interpreted as uniform random variables (2.22), and their constant density functions given with (4.57). Assuming the statistical independence of the elements $c_{m,q=1,\ldots,6}$, the density functions for the background color feature vectors $\boldsymbol{c}_m \notin C_O$ can be written according to (2.21) as

$$p(\boldsymbol{c}_m) = \prod_{q=1}^{6} \frac{1}{\max(c_{m,q}) - \min(c_{m,q})} = p_b \quad , \tag{4.69}$$

where p_b is a constant value. All pixels in RGB images are represented by the red, the green, and the blue value belonging to the set $\{0, 1, \ldots, 255\}$, and the color feature vectors computed with (4.62) using the Johnston wavelet. Thus, the background density p_b for the color modeling amounts approximately to $e^{-10.5}$. This value is used in the recognition phase for the object classification and localization in the real world environment.

4.3.5 Summary

The object and the background modeling using color feature vectors presented in Section 4.3 result in statistical object models $\mathcal{M} = \{\mathcal{M}_1, \mathcal{M}_2, \ldots, \mathcal{M}_\kappa, \ldots, \mathcal{M}_{N_\Omega}\}$ for color images, which are created for all object classes $\Omega = \{\Omega_1, \Omega_2, \ldots, \Omega_\kappa, \ldots, \Omega_{N_\Omega}\}$ considered in the recognition task. The object models $\mathcal{M}_\kappa$ are treated as continuous functions

$$\mathcal{M}_\kappa = \mathcal{M}_\kappa(\boldsymbol{\phi}, \boldsymbol{t}) \tag{4.70}$$

defined on the transformation parameter domain $(\boldsymbol{\phi}, \boldsymbol{t})$. Using these object models $\mathcal{M}_\kappa$ the object area

$$O_\kappa = O_\kappa(\boldsymbol{\phi}, \boldsymbol{t}) \tag{4.71}$$

can be determined for all objects Ω_κ in all poses $(\boldsymbol{\phi}, \boldsymbol{t})$. The six-dimensional local feature vectors extracted on the grid points $\boldsymbol{x}_{\kappa,m} \in O_\kappa(\boldsymbol{\phi}, \boldsymbol{t})$ are represented in the object models $\mathcal{M}_\kappa$ by normal density functions with the mean vectors $\boldsymbol{\mu}_{\kappa,m}(\boldsymbol{\phi}, \boldsymbol{t})$ and the standard deviation vectors $\boldsymbol{\sigma}_{\kappa,m}$ as follows

$$\boldsymbol{\mu}_{\kappa,m} = \boldsymbol{\mu}_{\kappa,m}(\boldsymbol{\phi}, \boldsymbol{t}) = \begin{pmatrix} \mu_{\kappa,m,1}(\boldsymbol{\phi}, \boldsymbol{t}) \\ \mu_{\kappa,m,2}(\boldsymbol{\phi}, \boldsymbol{t}) \\ \mu_{\kappa,m,3}(\boldsymbol{\phi}, \boldsymbol{t}) \\ \mu_{\kappa,m,4}(\boldsymbol{\phi}, \boldsymbol{t}) \\ \mu_{\kappa,m,5}(\boldsymbol{\phi}, \boldsymbol{t}) \\ \mu_{\kappa,m,6}(\boldsymbol{\phi}, \boldsymbol{t}) \end{pmatrix} , \quad \boldsymbol{\sigma}_{\kappa,m} = \begin{pmatrix} \sigma_{\kappa,m,1} \\ \sigma_{\kappa,m,2} \\ \sigma_{\kappa,m,3} \\ \sigma_{\kappa,m,4} \\ \sigma_{\kappa,m,5} \\ \sigma_{\kappa,m,6} \end{pmatrix} . \tag{4.72}$$

Additionally, each object model $\mathcal{M}_\kappa$ stores the background density value p_b.

4.4 Context Modeling

Section 3.1.5 presents the general idea of object modeling and recognition. The statistical algorithms for object classification and localization maximize the object density (3.24) over all possible object classes $\Omega = \{\Omega_1, \Omega_2, \ldots, \Omega_\kappa, \ldots, \Omega_{N_\Omega}\}$ and all pose hypotheses $(\boldsymbol{\phi}, \boldsymbol{t})$. The maximum likelihood estimation (ML) (3.24) is based on the assumption that the probability of the object occurrence in a scene is uniformly distributed (2.22) over all object classes $\Omega_{\kappa=1,\ldots,N_\Omega}$ (3.21)

$$p(\Omega_1) = \cdots = p(\Omega_\kappa) = \cdots = p(\Omega_{N_\Omega}) = \frac{1}{N_\Omega} . \tag{4.73}$$

Figure 4.8: Left: office context. Middle: kitchen context. Right: nursery context.

However, having additional knowledge about the environment, in which a scene was taken, the occurrence of some objects might be more likely than the occurrence of the others. Considering this additional knowledge in the learning phase is called *context modeling*. Figure 4.8 shows three example contexts, namely the office context, the kitchen context, and the nursery context. In the office context, objects like punchers, staplers, or pens can be found more likely than, e. g., plates, knifes, or forks, which are rather found in the kitchen. Therefore, it is useful to model the context dependencies between the objects in the learning phase. This is done by training of the a-priori probabilities for all object classes Ω_κ considered in the recognition task for each context separately.

First, the set Υ of contexts $\Upsilon_{\iota=1,\ldots,N_\Upsilon}$ in a particular object recognition task is introduced

$$\Upsilon = \{\Upsilon_1, \Upsilon_2, \ldots, \Upsilon_\iota, \ldots, \Upsilon_{N_\Upsilon}\} \quad . \tag{4.74}$$

It is assumed that the number N_Υ and the kinds[5] of the contexts are known. Moreover, the set of object classes $\Omega = \{\Omega_1, \Omega_2, \ldots, \Omega_\kappa, \ldots, \Omega_{N_\Omega}\}$ is also known for the learning of the context dependencies. The training of the context dependencies between objects starts with the image acquisition (Section 3.1.1). First, N_ι images from random viewpoints are taken with a hand-held camera (see Figure 4.3, left) for each context Υ_ι. Second, it is manually counted, which of the objects $\Omega_{\kappa=1,\ldots,N_\Omega}$ and how often occur in the images, whereas with $N_{\iota,\kappa}$ the number is denoted, how often the object Ω_κ occurs in the context Υ_ι. Generally, the sum of $N_{\iota,\kappa}$ for all object classes $\Omega_{\kappa=1,\ldots,N_\Omega}$ is not equal to N_ι

$$\frac{1}{N_\iota}\left(N_{\iota,1} + N_{\iota,2} + \ldots + N_{\iota,\kappa} + \ldots + N_{\iota,N_\Omega}\right) \neq 1 \quad . \tag{4.75}$$

[5]By context kinds environments like office, kitchen, nursery, bathroom etc. are meant.

Therefore, for all contexts $\Upsilon_{\iota=1,\ldots,N_\Upsilon}$ a normalization factor η_ι is defined as follows

$$\eta_\iota = \frac{N_\iota}{\sum_{\kappa=1}^{N_\Omega} N_{\iota,\kappa}} \quad . \tag{4.76}$$

Using this normalization factor η_ι and the number $N_{\iota,\kappa}$, the a-priori occurrence probability for the object Ω_κ in the context Υ_ι is learned as

$$p_\iota(\Omega_\kappa) = \eta_\iota N_{\iota,\kappa} \quad . \tag{4.77}$$

It means that for the recognition phase dealing with multi-object scenes under consideration of context dependencies (Section 5.5) the maximum likelihood estimation (ML) (3.24) cannot be used in this form. The nonuniform object occurrence probability distribution (4.77) learned for each context Υ_ι in the present section should be taken into account.

The contexts are trained separately from the objects (see Figure 4.1). For all regarded contexts $\Upsilon_{\iota=1,\ldots,N_\Upsilon}$ statistical context models $\mathcal{M}_{\iota=1,\ldots,N_\Upsilon}$ are learned. These context models contain a-priori densities $p_\iota(\Omega_\kappa)$ for all objects classes $\Omega_{\kappa=1,\ldots,N_\Omega}$ taken into account in the recognition task. Chapter 5.5 shows that the object classification and localization with the context modeling requires not only the context models $\mathcal{M}_\iota$, but also the statistical object models $\mathcal{M}_\kappa(\boldsymbol{\phi}, \boldsymbol{t})$ created using either gray level images (Section 4.2) or color images (Section 4.3).

4.5 Summary

This chapter described the training phase of the object recognition system. In the introduction to this chapter, a general description of all learning steps, which result in statistical object models and, if considered, in a context model, was presented and depicted in Figure 4.1.

Section 4.1 dealt with the problem of training data collection. Under training data for the object modeling, both the images $\boldsymbol{f}_{\kappa,\rho=1,\ldots,N_\rho}$ of the objects and the object poses $(\boldsymbol{\phi}_{\kappa,\rho}, \boldsymbol{t}_{\kappa,\rho})$ in these images are understood; while by training data for the context modeling, only the images for all contexts $\Upsilon_{\iota=1,\ldots,N_\Upsilon}$ taken with a hand-held camera are meant. Recording objects with a special setup like a turntable and camera arm (see Figure 4.2) yields both the images and the object poses (Section 4.1.1). After taking images with a hand-held camera (see Figure 4.3, left), the object poses are unknown. They are reconstructed using the structure-from-motion algorithm, which was presented in Section 4.1.2.

The statistical object modeling using gray level images was explained in Section 4.2. First, the original training RGB images are converted and resized into gray level images of size $2^n \times 2^n$ ($n \in \mathbb{N}$) pixels. Second, 2D local feature vectors $\boldsymbol{c}_{\kappa,m}$ are extracted in these images using the wavelet transformation. The feature vectors computation was described in detail in Section 4.2.1. Section 4.2.2 defined the object area, i. e., the set of grid points belonging to the object. This object area can be regarded as a function $O_\kappa(\boldsymbol{\phi}, \boldsymbol{t})$ defined for all object classes Ω_κ on a continuous pose parameter domain $(\boldsymbol{\phi}, \boldsymbol{t})$. For each object class Ω_κ and pose parameters $(\boldsymbol{\phi}, \boldsymbol{t})$, it determines the set C_{O_κ} of feature vectors $\boldsymbol{c}_{\kappa,m}$ describing the object. The remaining feature vectors $\boldsymbol{c}_{\kappa,m} \notin C_{O_\kappa}$ are called background features. The object feature vectors $\boldsymbol{c}_{\kappa,m} \in C_{O_\kappa}$ are modeled with the normal density functions, which was presented in Section 4.2.3, whereas the corresponding mean value vectors $\boldsymbol{\mu}_{\kappa,m}$ are represented as functions of the external pose parameters $\boldsymbol{\mu}_{\kappa,m}(\phi_{\text{ext}}, t_{\text{ext}})$ and the standard deviation vectors $\boldsymbol{\sigma}_{\kappa,m}$ are modeled as constant. The background feature vectors $\boldsymbol{c}_{\kappa,m} \notin C_{O_\kappa}$ are modeled with the uniform density functions. Their value is constant for all background features and denoted by p_b (Section 4.2.4). Section 4.2.5 summarized all components of the statistical object models $\mathcal{M}_\kappa(\boldsymbol{\phi}, \boldsymbol{t})$ for gray level images, which are created for all object classes Ω_κ and considered as continuous functions of the transformation parameters $(\boldsymbol{\phi}, \boldsymbol{t})$.

Section 4.3 presented the color modeling, i. e., the statistical object modeling for color images. Generally, it consists of the same steps as the object modeling for gray level images, but it is different in object description with feature vectors. After image resizing into RGB images of size $2^n \times 2^n$ ($n \in \mathbb{N}$) pixels, 6D local feature vectors $\boldsymbol{c}_{\kappa,m}$ are computed with the wavelet transformation. The first $c_{\kappa,m,1}$ and the second $c_{\kappa,m,1}$ element of the feature vector $\boldsymbol{c}_{\kappa,m}$ result from the red channel, the third $c_{\kappa,m,3}$ and fourth $c_{\kappa,m,4}$ depend on the green channel, and the fifth $c_{\kappa,m,5}$ and the sixth $c_{\kappa,m,6}$ are computed from the blue channel (Section 4.3.1). Analogical to the object modeling using gray level images, Section 4.3.2 defined the object area $O_\kappa(\boldsymbol{\phi}, \boldsymbol{t})$ for the color modeling. Statistical modeling of object feature vectors with normal density functions was described in Section 4.3.3, while Section 4.3.4 modeled the background features with uniform density functions. All components of the statistical object models $\mathcal{M}_\kappa(\boldsymbol{\phi}, \boldsymbol{t})$ for color images were summarized in Section 4.3.5.

In known contexts the occurrence of some objects is more likely than the occurrence of the others. In this case, it is useful to learn the a-priori probabilities $p_\iota(\Omega_\kappa)$ for all object classes Ω_κ in all known contexts Υ_ι, which was discussed in Section 4.4. The context models $\mathcal{M}_\iota$ are trained separately from the object models $\mathcal{M}_\kappa(\boldsymbol{\phi}, \boldsymbol{t})$ and contain the learned densities $p_\iota(\Omega_\kappa)$ for all object classes Ω_κ considered in the object recognition task.

Chapter 5

Classification and Localization

A system for object recognition works in two different modes (see Figure 1.3). Since for all object classes $\Omega_{\kappa=1,\ldots,N_\Omega}$ regarded in a particular recognition task corresponding statistical object models $\mathcal{M}_{\kappa=1,\ldots,N_\Omega}$ and, if considered, for all contexts $\Upsilon_{\iota=1,\ldots,N_\Upsilon}$ corresponding statistical context models $\mathcal{M}_{\iota=1,\ldots,N_\Upsilon}$ are already trained (Chapter 4), the system is able to classify and localize objects in images taken from a real world environment [Grz05b]. As mentioned in Section 1.1, the recognition algorithm determines not only the classes of objects which occur in the image, but also their poses in terms of object pose definition illustrated in Figure 1.2. In this chapter all algorithms for object classification and localization which are integrated into the system, are discussed. Figure 5.1 illustrates the general scheme of the recognition phase.

First, a test image $\boldsymbol{f}$ from a real world environment is taken, preprocessed, and local feature vectors in it are computed in the same way as in the training phase [Grz05a]. The image preprocessing and feature extraction are briefly reviewed in Section 5.1. After that, one of the four recognition algorithms is manually chosen and started, whereas the system assumes that at least one of the objects $\Omega = \{\Omega_1, \Omega_2, \ldots, \Omega_\kappa, \ldots, \Omega_{N_\Omega}\}$ occurs in the test image $\boldsymbol{f}$. The recognition algorithm dealing with single-object scenes and using a single view of an object is presented in Section 5.2. Section 5.3 concerns the object classification and localization algorithm for single-object scenes which uses multiple views of an object. The system is also able to determine object classes and poses in multi-object scenes. The description of the recognition algorithm for multi-object scenes, which does not take into consideration any context dependencies, follows in Section 5.4. The classification and localization algorithm for multi-object scenes with context dependencies, which uses the context models $\mathcal{M}_\iota$ learned in the training phase, is presented in Section 5.5. Section 5.6 closes this chapter by summarizing the whole recognition phase of the system.

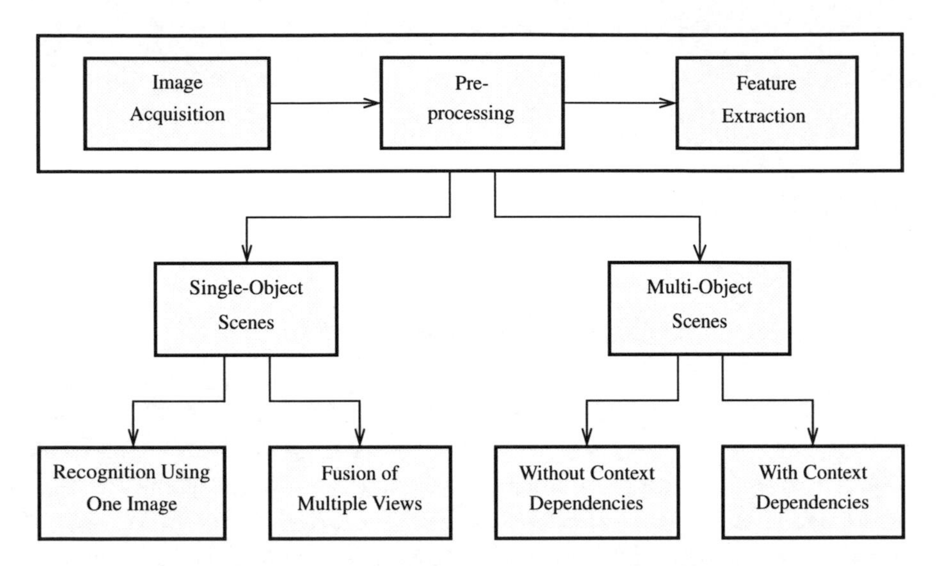

Figure 5.1: General scheme of the recognition phase. First, an image from a real world environment is taken, preprocessed, and feature vectors in it are computed. Then the system starts one of the four recognition algorithms. The first algorithm deals with single-object scenes using one image, the second one works also with single-object scenes, but it uses multiple views of the object, the third one is applied for multi-object scenes without context dependencies, and the fourth one considers context dependencies in the multi-object scenes.

5.1 Image Preprocessing and Feature Extraction

The preprocessing of the test image $\boldsymbol{f}$ and the feature extraction in the preprocessed images is illustrated in Figure 5.2. If object models for gray level images (Section 4.2) are used, then the original test image $\boldsymbol{f}$ is preprocessed into square gray level image $\boldsymbol{f}_g$ of the same size ($2^n \times 2^n$ pixels) as images in the training phase [Grz05b]. After that a set

$$C_g = \{\boldsymbol{c}_{g,1}, \boldsymbol{c}_{g,2}, \ldots, \boldsymbol{c}_{g,\widehat{m}}, \ldots, \boldsymbol{c}_{g,\widehat{M}}\} \tag{5.1}$$

of two-dimensional local feature vectors $\boldsymbol{c}_{g,\widehat{m}}$ is determined with the wavelet transformation [Mal89] in the same way as described in Section 4.2.1. The number $\widehat{M}$ of all feature vectors is given by

$$\widehat{M} = \frac{2^n \cdot 2^n}{2^{|\widehat{s}|} \cdot 2^{|\widehat{s}|}} \quad , \tag{5.2}$$

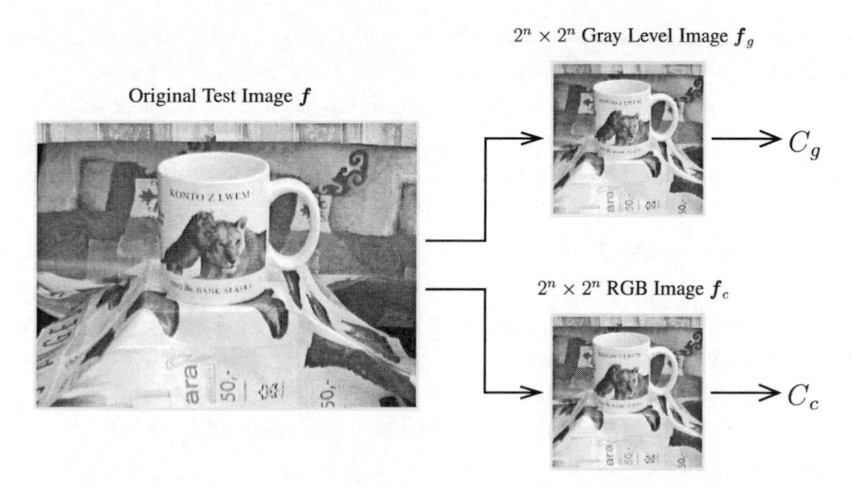

Figure 5.2: Image preprocessing and feature extraction in the recognition phase. The original test image $\boldsymbol{f}$ can be preprocessed into a square gray level image $\boldsymbol{f}_g$ or a square color image $\boldsymbol{f}_c$. In the gray level image a set of 2D local feature vectors C_g, while in the color image a set of 6D local feature vectors C_c is determined. Performing the image acquisition and preprocessing, it has to be made sure that the objects are not cut during this process.

where $|\widehat{s}|$ is the scale coefficient of the wavelet transformation. The local feature vectors are computed from square neighborhoods of size $2^{|\widehat{s}|} \cdot 2^{|\widehat{s}|}$ pixels, i. e., $|\widehat{s}|$ steps of the wavelet transformation have to be performed in order to compute the feature vector $\boldsymbol{c}_{g,\widehat{m}}$. Figure 4.4 illustrates the feature extraction from gray level images for the scale $|\widehat{s}| = 2$ of the wavelet signal decomposition. The feature vectors for this scale are given with the formula (4.16). The general definition of a gray level feature vector for any scale $|\widehat{s}|$ can be found in the equation (4.17) in Section 4.2.1.

The system is also able to train object models using color images (Section 4.3). In this case in the recognition phase the original test image $\boldsymbol{f}$ is preprocessed into square RGB image $\boldsymbol{f}_c$ (see Figure 5.2) of the same size ($2^n \times 2^n$ pixels) as images in the training phase. Then six-dimensional local feature vectors $\boldsymbol{c}_{c,\widehat{m}}$ are computed from the preprocessed test image $\boldsymbol{f}_c$ [Grz05a] and can be written together as a set

$$C_c = \{\boldsymbol{c}_{c,1}, \boldsymbol{c}_{c,2}, \ldots, \boldsymbol{c}_{c,\widehat{m}}, \ldots, \boldsymbol{c}_{c,\widehat{M}}\} \quad . \tag{5.3}$$

The number $\widehat{M}$ of color feature vectors is equal to the number or gray level feature vectors, i. e., it is given with the equation (5.2). The feature vectors $\boldsymbol{c}_{c,\widehat{m}}$ are computed with the 2D discrete signal decomposition with the wavelet transformation [Mal89], which is done separately for the red, the green, and the blue channel [Grz05a]. The computation of the wavelet coefficients for color feature extraction is illustrated in Figure 4.7, whereas the wavelet analysis scale amounts to $|\widehat{s}| = 2$. The six-dimensional local feature vector $\boldsymbol{c}_{c,\widehat{m}}$ for any resolution scale $|\widehat{s}|$ of the wavelet signal analysis is given with the equation (4.62) in Section 4.3.1.

Further mathematical description following in this chapter is identical for the recognition algorithms using gray level images and color images. Therefore, both the set C_g of gray level feature vectors $\boldsymbol{c}_{g,\widehat{m}}$ as well as the set C_c of color feature vectors $\boldsymbol{c}_{c,\widehat{m}}$ are denoted in the same way, namely as C. Moreover, the gray level feature vectors $\boldsymbol{c}_{g,\widehat{m}}$ and the color feature vectors $\boldsymbol{c}_{c,\widehat{m}}$, which are computed from the preprocessed test images in the recognition phase, are also written consistently as $\boldsymbol{c}_{\widehat{m}}$, while the preprocessed test images $\boldsymbol{f}_g$ and $\boldsymbol{f}_c$ are denoted with $\boldsymbol{f}$.

5.2 Single-Object Scenes

The algorithm described in the present section uses only one image $\boldsymbol{f}$ (one object view) for object classification and localization. Besides, it is assumed that in the scene $\boldsymbol{f}$ exactly one object from the set of objects $\Omega = \{\Omega_1, \Omega_2, \ldots, \Omega_\kappa, \ldots, \Omega_{N_\Omega}\}$ learned in the training phase occurs. The object models $\mathcal{M} = \{\mathcal{M}_1, \mathcal{M}_2, \ldots, \mathcal{M}_\kappa, \ldots, \mathcal{M}_{N_\Omega}\}$ for all considered object classes as well as the set C of local feature vectors $\boldsymbol{c}_{\widehat{m}}$ computed from the preprocessed test scene $\boldsymbol{f}$ are already known. The statistical object models were created in the training phase (Chapter 4), while the local feature vectors $\boldsymbol{c}_{\widehat{m}}$ observed in the preprocessed test image $\boldsymbol{f}$ (Section 5.1). The task of the classification and localization algorithm for single-object scenes is to find the class $\Omega_{\widehat{\kappa}}$, (or just its index $\widehat{\kappa}$) and the pose $(\widehat{\boldsymbol{\phi}}, \widehat{\boldsymbol{t}})$ of the object, which occurs in the test image $\boldsymbol{f}$.

In order to classify (determine the class number $\widehat{\kappa}$) and localize (determine the object pose $(\widehat{\boldsymbol{\phi}}, \widehat{\boldsymbol{t}})$) the object in the scene $\boldsymbol{f}$ the object density values for all objects Ω_κ and many pose hypotheses $(\boldsymbol{\phi}_h, \boldsymbol{t}_h)$ have to be compared to each other. Section 5.2.1 describes how to compute the object density value for the given test image $\boldsymbol{f}$, object class Ω_κ, and object pose hypothesis $(\boldsymbol{\phi}_h, \boldsymbol{t}_h)$. The recognition algorithm based on the maximum likelihood estimation is presented in Section 5.2.2. Section 5.2.3 extends this recognition algorithm. The extended algorithm uses object features extracted on different resolution levels of the wavelet transformation.

5.2.1 Object Density Value

In this section, the computation of the object density value for the given test image $\boldsymbol{f}$, object class hypothesis Ω_κ, and object pose hypothesis $(\boldsymbol{\phi}_h, \boldsymbol{t}_h)$ is explained. After the training phase described in Chapter 4, the object model $\mathcal{M}_\kappa(\boldsymbol{\phi}, \boldsymbol{t})$ for the object class Ω_κ is known. However, there are two possibilities to create the statistical object models. First, the training phase can be performed using gray level images (Section 4.2). Second, RGB images can be used for learning of the object models (Section 4.3). In the case of gray level modeling, the object model $\mathcal{M}_\kappa(\boldsymbol{\phi}, \boldsymbol{t})$ contains components summarized in Section 4.2.5 and the objects are described with two-dimensional local feature vectors (vector dimension $N_q = 2$). Components of the color object model $\mathcal{M}_\kappa(\boldsymbol{\phi}, \boldsymbol{t})$ are listed in Section 4.3.5. In this case, objects are represented by six-dimensional local feature vectors (vector dimension $N_q = 6$). Using an additional parameter N_q, which denotes the feature vector dimension, the density functions (4.41) and (4.68) can be expressed together with[1]

$$p(C_{O_\kappa}|\boldsymbol{B}_\kappa, \boldsymbol{\phi}, \boldsymbol{t}) = \prod_{\boldsymbol{x}_m \in O_\kappa} \prod_{q=1}^{N_q} p(c_{m,q}|\mu_{\kappa,m,q}, \sigma_{\kappa,m,q}, \boldsymbol{\phi}, \boldsymbol{t}) \quad . \tag{5.4}$$

Therefore, further considerations in this section regard both the algorithm for object classification and localization for gray level and color images.

Let's assume, the test image $\boldsymbol{f}$ is taken, preprocessed, and the set C of local feature vectors $\boldsymbol{c}_{\widehat{m}}$ is determined in it according to Section 5.1. Besides, the object class hypothesis Ω_κ and the object pose hypothesis $(\boldsymbol{\phi}_h, \boldsymbol{t}_h)$ are given. The computation of the object density value $p(C_{O_\kappa}|\boldsymbol{B}_\kappa, \boldsymbol{\phi}_h, \boldsymbol{t}_h)$ is done in the following way:

1. *Object Area* $O_\kappa(\boldsymbol{\phi}_h, \boldsymbol{t}_h)$

   In the object model $\mathcal{M}_\kappa$ the set of grid points $\boldsymbol{x}_m$, which belong to the object, is stored. This set is called object area and is trained as a continuous function of the object pose parameters (4.60) and (4.71)

$$O_\kappa = O_\kappa(\boldsymbol{\phi}, \boldsymbol{t}) \quad . \tag{5.5}$$

   Thus, the object area O_κ can be also determined for the hypothesis object pose $(\boldsymbol{\phi}_h, \boldsymbol{t}_h)$. Only feature vectors $\boldsymbol{c}_m$ computed on the grid points $\boldsymbol{x}_m \in O_\kappa(\boldsymbol{\phi}_h, \boldsymbol{t}_h)$ belonging to the object area should be taken into consideration for the calculation of the object density value. The training of the object area for gray level images is described in detail in Section 4.2.2, while the object area for color images is defined in Section 4.3.2.

[1] Remember that in Chapter 4 the class index κ was omitted except Sections 4.2.5, 4.3.5, and 4.5.

2. *Bilinear Interpolation of Feature Vectors* $\boldsymbol{c}_m$
   Figure 4.5 explains the difference between the object grid and the image grid. In Section 5.1 the N_q-dimensional local feature vectors $\boldsymbol{c}_{\widehat{m}}$ are extracted on the image grid points $\boldsymbol{x}_{\widehat{m}}$. However, for the density value computation feature vectors $\boldsymbol{c}_m$ determined on the object grid points $\boldsymbol{x}_m$ are required. If the object is transformed with the internal pose parameters $(\phi_{\text{int},h}, \boldsymbol{t}_{\text{int},h})$ its grid points $\boldsymbol{x}_m$, in general, do not coincide with the positions $\boldsymbol{x}_{\widehat{m}}$ of the image grid. In this case the object feature vectors $\boldsymbol{c}_m$ on the transformed positions $\boldsymbol{x}_m$ are bilinear interpolated from the four adjacent image feature vectors, which is depicted in Figure 4.6.

3. *Set of Object Feature Vectors* $C_{O_\kappa}(\boldsymbol{\phi}_{\text{ext},h}, \boldsymbol{t}_{\text{ext},h})$
   According to Section 4.2.2 (4.36) and Section 4.3.2 the set of object feature vectors C_{O_κ} as a continuous function of the external pose parameters $(\boldsymbol{\phi}_{\text{ext}}, \boldsymbol{t}_{\text{ext}})$ is also stored in the object model $\mathcal{M}_\kappa$

$$C_{O_\kappa} = C_{O_\kappa}(\boldsymbol{\phi}_{\text{ext}}, \boldsymbol{t}_{\text{ext}}) \quad . \tag{5.6}$$

   Thus, it can be determined for the hypothesis external pose parameters $(\boldsymbol{\phi}_{\text{ext},h}, \boldsymbol{t}_{\text{ext},h})$. For the computation of the object density value only feature vectors belonging to this set, i. e., $\boldsymbol{c}_m \in C_{O_\kappa}(\boldsymbol{\phi}_{\text{ext},h}, \boldsymbol{t}_{\text{ext},h})$, should be taken into account. Other feature vectors $\boldsymbol{c}_m \notin C_{O_\kappa}(\boldsymbol{\phi}_{\text{ext},h}, \boldsymbol{t}_{\text{ext},h})$ represent the background, which is a-priori unknown in the recognition phase, and must not influence the object density value.

4. *Density Values for Object Feature Vectors* $p(\boldsymbol{c}_m|\boldsymbol{\mu}_{\kappa,m}, \boldsymbol{\sigma}_{\kappa,m}, \boldsymbol{\phi}_h, \boldsymbol{t}_h)$
   As described in Section 4.2.3 (4.39) and in Section 4.3.3 (4.67) for all object feature vectors $\boldsymbol{c}_m \in C_{O_\kappa}(\boldsymbol{\phi}_{\text{ext},h}, \boldsymbol{t}_{\text{ext},h})$ corresponding normal density functions with mean vectors $\boldsymbol{\mu}_{\kappa,m}$ and standard deviation vectors $\boldsymbol{\sigma}_{\kappa,m}$ are modeled in the training phase and stored in the statistical object model $\mathcal{M}_\kappa$. Therefore, the density values $p(\boldsymbol{c}_m|\boldsymbol{\mu}_{\kappa,m}, \boldsymbol{\sigma}_{\kappa,m}, \boldsymbol{\phi}_h, \boldsymbol{t}_h)$ for the object feature vectors $\boldsymbol{c}_m \in C_{O_\kappa}(\boldsymbol{\phi}_{\text{ext},h}, \boldsymbol{t}_{\text{ext},h})$ observed in the test image $\boldsymbol{f}$ can be easily determined. The better the object feature vector $\boldsymbol{c}_m$ observed in the test image $\boldsymbol{f}$ matches the corresponding feature vector learned in the training phase (represented by $\boldsymbol{\mu}_{\kappa,m}$ and $\boldsymbol{\sigma}_{\kappa,m}$), the higher the density value $p(\boldsymbol{c}_m|\boldsymbol{\mu}_{\kappa,m}, \boldsymbol{\sigma}_{\kappa,m}, \boldsymbol{\phi}_h, \boldsymbol{t}_h)$ is for this feature.

5. *Object Density Value* $p(C_{O_\kappa}|\boldsymbol{B}_\kappa, \boldsymbol{\phi}_h, \boldsymbol{t}_h)$
   Assuming that the object feature vectors $\boldsymbol{c}_m \in C_{O_\kappa}$ are statistically independent on each other, the object density value $p(C_{O_\kappa}|\boldsymbol{B}_\kappa, \boldsymbol{\phi}_h, \boldsymbol{t}_h)$ for the given test image $\boldsymbol{f}$, object class

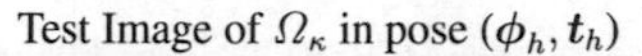

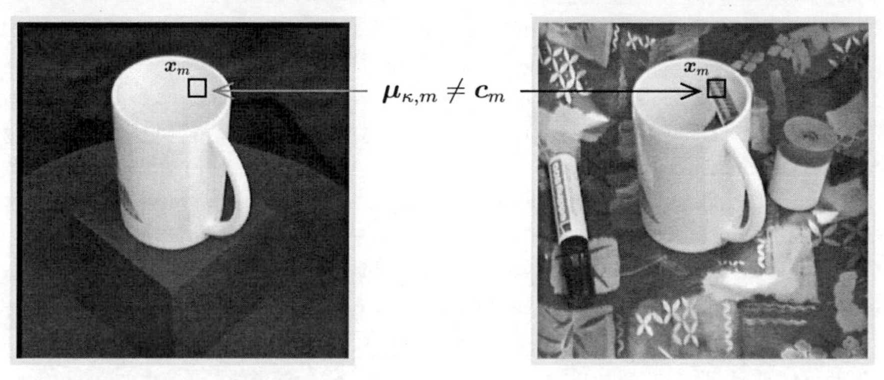

Figure 5.3: Training image and test image of the same object in the same pose. Due to the occlusion in the test image, the test feature vector c_m is totally different from the corresponding training feature represented by $\mu_{\kappa,m}$ and $\sigma_{\kappa,m}$. Thus, the density value for c_m is near to zero $p(c_m|\mu_{\kappa,m}, \sigma_{\kappa,m}, \phi_h, t_h) \approx 0$.

hypothesis Ω_κ, and object pose hypothesis (ϕ_h, t_h) can be computed with

$$p(C_{O_\kappa}|B_\kappa, \phi_h, t_h) = \prod_{c_m \in C_{O_\kappa}} p(c_m|\mu_{\kappa,m}, \sigma_{\kappa,m}, \phi_h, t_h) \quad . \tag{5.7}$$

In the ideal case the better the object Ω_κ in the pose (ϕ_h, t_h) matches the object in the test image f represented by the set C_{O_κ} of local feature vectors, the higher the object density value $p(C_{O_\kappa}|B_\kappa, \phi_h, t_h)$ is.

In Figure 5.3 a training image of the object Ω_κ in the pose (ϕ_h, t_h) and a test image of the same object in the same pose taken from a real world environment are shown. One expects that the object density $p(C_{O_\kappa}|B_\kappa, \phi_h, t_h)$ (5.7) for this hypothesis $(\Omega_\kappa, \phi_h, t_h)$ and this test image f takes a relatively high value. However, if a feature vector c_m extracted from the test image f (test feature vector) is totally different from the corresponding training feature vector represented by $\mu_{\kappa,m}$ and $\sigma_{\kappa,m}$, the density value for this feature vector is near to zero $p(c_m|\mu_{\kappa,m}, \sigma_{\kappa,m}, \phi_h, t_h) \approx 0$. Hence, the value of the object density (5.7) is also near to zero

$$(\exists\, c_m \in C_{O_\kappa} : p(c_m|\mu_{\kappa,m}, \sigma_{\kappa,m}, \phi_h, t_h) \approx 0) \quad \Rightarrow \quad p(C_{O_\kappa}|B_\kappa, \phi_h, t_h) \approx 0 \quad . \tag{5.8}$$

Since the heterogeneous background in the recognition phase is a-priori unknown and, in general, different from the background in the training phase, the test features $\boldsymbol{c}_m$ observed on the border of the object area $O_\kappa(\boldsymbol{\phi}_h, \boldsymbol{t}_h)$ differ significantly from the corresponding training feature vectors represented by $\boldsymbol{\mu}_{\kappa,m}$ and $\boldsymbol{\sigma}_{\kappa,m}$. The same situation occurs for feature vectors $\boldsymbol{c}_m$ extracted on occlusions (see Figure 5.3), which are also a-priori unknown and cannot be modeled in the training phase. Therefore, the comparison of the object density values computed according to (5.7) for different object classes Ω_κ and pose hypotheses $(\boldsymbol{\phi}_h, \boldsymbol{t}_h)$ does not yield satisfying classification and localization rates for scenes with heterogeneous background and occlusions.

In order to solve the problem of object classification and localization in a real world environment, the equation (5.7) for computation of the object density value has to be modified by considering the background density p_b trained in Section 4.2.4 and Section 4.3.4. In other words, for each object feature vector $\boldsymbol{c}_m \in C_{O_\kappa}(\boldsymbol{\phi}_{\text{ext},h}, t_{\text{ext},h})$ extracted from the test image $\boldsymbol{f}$ it has to be revised, if it really describes the object, or belongs to the background and the occlusion respectively. This revision is done with the following rule

$$\forall\ \boldsymbol{c}_m \in C_{O_\kappa}: \quad p(\boldsymbol{c}_m | \boldsymbol{\mu}_{\kappa,m}, \boldsymbol{\sigma}_{\kappa,m}, \boldsymbol{\phi}_h, \boldsymbol{t}_h) < p_b \quad \Rightarrow \quad \boldsymbol{c}_m \notin C'_{O_\kappa} \quad , \tag{5.9}$$

where $C'_{O_\kappa} \subset C_{O_\kappa}$ denotes the set of local feature vectors $\boldsymbol{c}_m$, which really describe the object in the test scene $\boldsymbol{f}$. Considering (5.9), the formula for the object density value (5.7) modifies to

$$p(C_{O_\kappa} | \widehat{\boldsymbol{B}}_\kappa, \boldsymbol{\phi}_h, \boldsymbol{t}_h) = \prod_{\boldsymbol{c}_m \in C_{O_\kappa}} \max\{p(\boldsymbol{c}_m | \boldsymbol{\mu}_{\kappa,m}, \boldsymbol{\sigma}_{\kappa,m}, \boldsymbol{\phi}_h, \boldsymbol{t}_h),\ p_b\} \quad . \tag{5.10}$$

$\widehat{\boldsymbol{B}}_\kappa$ contains not only the learned statistical parameters of the object feature vectors ($\boldsymbol{\mu}_{\kappa,m}$ and $\boldsymbol{\sigma}_{\kappa,m}$) like $\boldsymbol{B}_\kappa$, but also the background density p_b defined by (4.58) in Section 4.2.4 and (4.69) in Section 4.3.4. Due to the computation of the object density value using (5.10), the object recognition problem for scenes with heterogeneous background and occlusions illustrated in Figure 5.3 and mathematically described by (5.8) can be solved.

5.2.2 Object Density Maximization

Section 5.2.1 explains in detail how to compute the object density value $p(C_{O_\kappa} | \widehat{\boldsymbol{B}}_\kappa, \boldsymbol{\phi}_h, \boldsymbol{t}_h)$ for the given test image $\boldsymbol{f}$, object class hypothesis Ω_κ, and object pose hypothesis $(\boldsymbol{\phi}_h, \boldsymbol{t}_h)$. However, the objective of object classification and localization for single-object scenes is to find the "best" object hypothesis $\Omega_{\widehat{\kappa}}$ and the "best" pose hypothesis $(\widehat{\boldsymbol{\phi}}, \widehat{\boldsymbol{t}})$ for the test image $\boldsymbol{f}$. It is still not proved that the object density value (5.10) is a good measure in terms of object classification and localization. Let's assume that $p(\Omega_\kappa, \boldsymbol{\phi}_h, \boldsymbol{t}_h | C_{O_\kappa})$ denotes the probability for the class Ω_κ and the pose $(\boldsymbol{\phi}_h, \boldsymbol{t}_h)$ in the test image $\boldsymbol{f}$ represented by the set of object feature vectors C_{O_κ}. Searching for the classification result $\widehat{\kappa}$ and the localization result $(\widehat{\boldsymbol{\phi}}, \widehat{\boldsymbol{t}})$ the density $p(\Omega_\kappa, \boldsymbol{\phi}_h, \boldsymbol{t}_h | C_{O_\kappa})$ is maximized, which can be expressed with the maximum a-posteriori estimation (MAP) introduced in Section 2.1 (2.28) as follows

$$(\widehat{\kappa}, \widehat{\boldsymbol{\phi}}, \widehat{\boldsymbol{t}}) = \underset{(\kappa, \boldsymbol{\phi}_h, \boldsymbol{t}_h)}{\operatorname{argmax}}\, p(\Omega_\kappa, \boldsymbol{\phi}_h, \boldsymbol{t}_h | C_{O_\kappa}) = \underset{(\kappa, \boldsymbol{\phi}_h, \boldsymbol{t}_h)}{\operatorname{argmax}} \{p(\Omega_\kappa, \boldsymbol{\phi}_h, \boldsymbol{t}_h) p(C_{O_\kappa} | \Omega_\kappa, \boldsymbol{\phi}_h, \boldsymbol{t}_h)\} \quad . \tag{5.11}$$

Since the density values $p(\Omega_\kappa, \boldsymbol{\phi}_h, \boldsymbol{t}_h)$ are assumed to be equal for all object classes Ω_κ and all object pose hypotheses $(\boldsymbol{\phi}_h, \boldsymbol{t}_h)$, the MAP estimation (5.11) can be modified to the maximum likelihood estimation (ML)

$$(\widehat{\kappa}, \widehat{\boldsymbol{\phi}}, \widehat{\boldsymbol{t}}) = \underset{(\kappa, \boldsymbol{\phi}_h, \boldsymbol{t}_h)}{\operatorname{argmax}}\, p(C_{O_\kappa} | \Omega_\kappa, \boldsymbol{\phi}_h, \boldsymbol{t}_h) \quad , \tag{5.12}$$

which is introduced in Section 3.1.5 (3.24). Each object class Ω_κ can be represented by $\widehat{\boldsymbol{B}}_\kappa$ containing the statistical parameters for the object $(\boldsymbol{\mu}_{\kappa,m}, \boldsymbol{\sigma}_{\kappa,m})$ and the background p_b. Thus, (5.12) can be written as

$$(\widehat{\kappa}, \widehat{\boldsymbol{\phi}}, \widehat{\boldsymbol{t}}) = \underset{(\kappa, \boldsymbol{\phi}_h, \boldsymbol{t}_h)}{\operatorname{argmax}}\, p(C_{O_\kappa} | \widehat{\boldsymbol{B}}_\kappa, \boldsymbol{\phi}_h, \boldsymbol{t}_h) \quad . \tag{5.13}$$

As can be seen, the object classification $(\widehat{\kappa})$ and localization $(\widehat{\boldsymbol{\phi}}, \widehat{\boldsymbol{t}})$ for single-object scenes can be solved by maximization of the object density $p(C_{O_\kappa} | \widehat{\boldsymbol{B}}_\kappa, \boldsymbol{\phi}_h, \boldsymbol{t}_h)$ over the class and the pose parameters $(\kappa, \boldsymbol{\phi}_h, \boldsymbol{t}_h)$. Section 5.2.1 explains how to determine the object density value $p(C_{O_\kappa} | \widehat{\boldsymbol{B}}_\kappa, \boldsymbol{\phi}_h, \boldsymbol{t}_h)$ (5.10) for the given test image $\boldsymbol{f}$, object class Ω_κ, and object pose $(\boldsymbol{\phi}_h, \boldsymbol{t}_h)$.

The object density value $p(C_{O_\kappa} | \widehat{\boldsymbol{B}}_\kappa, \boldsymbol{\phi}_h, \boldsymbol{t}_h)$ determined in Section 5.2.1 (5.10) depends not only on the test image $\boldsymbol{f}$, object class Ω_κ, and object pose hypothesis $(\boldsymbol{\phi}_h, \boldsymbol{t}_h)$, but also on the number N_κ of object feature vectors $\boldsymbol{c}_m \in C_{O_\kappa}$. Similarly to the set C_{O_κ}, the number of object feature vectors varies with the external pose parameters

$$N_\kappa = N_\kappa(\boldsymbol{\phi}_{\text{ext}}, t_{\text{ext}}) \quad . \tag{5.14}$$

Therefore, it can be determined for the hypothesis pose parameters $N_\kappa(\boldsymbol{\phi}_h, t_h)$. Since the most object feature densities $p(\boldsymbol{c}_m | \boldsymbol{\mu}_{\kappa,m}, \boldsymbol{\sigma}_{\kappa,m}, \boldsymbol{\phi}_h, \boldsymbol{t}_h)$ and the background density p_b take values between 0 and 1, the comparison (5.13) of the object density values computed by (5.10) privileges small objects with low number N_κ of object feature vectors. In order to decrease the influence of the number N_κ of object features on the classification ($\widehat{\kappa}$) and localization ($\widehat{\boldsymbol{\phi}}, \widehat{\boldsymbol{t}}$) result (5.13), the object density $p(C_{O_\kappa}|\widehat{\boldsymbol{B}}_\kappa, \boldsymbol{\phi}_h, \boldsymbol{t}_h)$ (5.10) is normalized by a *quality measure* Q

$$Q(p(C_{O_\kappa}|\widehat{\boldsymbol{B}}_\kappa, \boldsymbol{\phi}_h, \boldsymbol{t}_h)) = \sqrt[N_\kappa]{p(C_{O_\kappa}|\widehat{\boldsymbol{B}}_\kappa, \boldsymbol{\phi}_h, \boldsymbol{t}_h)} \quad . \tag{5.15}$$

The quality measure Q represents the geometric mean of the object feature density values and is called *geometric criterion*. Using the quality measure Q the maximization term (5.13) modifies to

$$(\widehat{\kappa}, \widehat{\boldsymbol{\phi}}, \widehat{\boldsymbol{t}}) = \underset{(\kappa, \boldsymbol{\phi}_h, \boldsymbol{t}_h)}{\operatorname{argmax}}\, Q(p(C_{O_\kappa}|\widehat{\boldsymbol{B}}_\kappa, \boldsymbol{\phi}_h, \boldsymbol{t}_h)) \quad . \tag{5.16}$$

The practical realization of the maximization algorithm (5.16) for object classification and localization is illustrated in Figure 5.4. First, the test image $\boldsymbol{f}$ is preprocessed to gray level image or color image. Second, feature vectors are computed in the preprocessed test image using the wavelet transformation. After that for each object class Ω_κ and each pose hypothesis $(\boldsymbol{\phi}_h, \boldsymbol{t}_h)$ the object area $O_\kappa(\boldsymbol{\phi}_h, \boldsymbol{t}_h)$ is determined. Hence, the set of object feature vectors $C_{O_\kappa}(\boldsymbol{\phi}_{\text{ext},h}, \boldsymbol{t}_{\text{ext},h})$ is also known. Then, according to Section 5.2.1, object density values

$$p_{\kappa,h} = p(C_{O_\kappa}|\widehat{\boldsymbol{B}}_\kappa, \boldsymbol{\phi}_h, \boldsymbol{t}_h) \tag{5.17}$$

are determined and normalized by the quality measure Q given by (5.15), which results in $Q(p_{\kappa,h})$. The final classification ($\widehat{\kappa}$) and localization ($\widehat{\boldsymbol{\phi}}, \widehat{\boldsymbol{t}}$) result corresponds to the highest value of the normalized density $Q(p_{\kappa,h})$.

5.2.3 Wavelet Multiresolution Analysis

Figure 4.4 and Figure 4.7 in Chapter 4 illustrate the local feature extraction in gray level and color images respectively for the wavelet resolution level $L_{\widehat{s}} = L_{-2}$, i.e., for local neighborhoods of size 4×4 pixels. According to (4.17) and (4.62) the feature extraction can be also performed for other values of the wavelet scale parameter $\widehat{s}$. In [Grz05b] the feature computation for three different wavelet resolution levels, namely L_{-1}, L_{-2}, and L_{-3}, is discussed. In this case for each object class Ω_κ three statistical object models $\mathcal{M}_{\kappa,\widehat{s}=-1,-2,-3}$

$$\Omega_\kappa \quad \longrightarrow \quad \begin{cases} \mathcal{M}_{\kappa,-1} \\ \mathcal{M}_{\kappa,-2} \\ \mathcal{M}_{\kappa,-3} \end{cases} \tag{5.18}$$

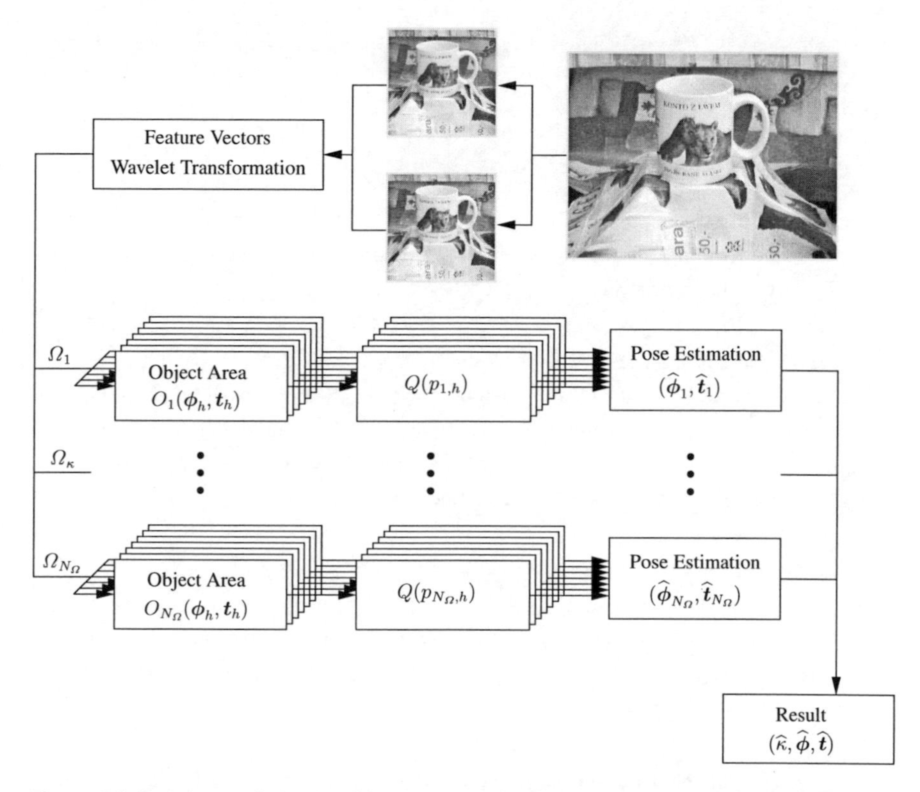

Figure 5.4: Density maximization for object classification and localization. First, local feature vectors from the preprocessed test image are computed. Then, for each object class Ω_κ and each pose hypothesis $(\boldsymbol{\phi}_h, \boldsymbol{t}_h)$ the object area $O_\kappa(\boldsymbol{\phi}_h, \boldsymbol{t}_h)$ is determined and the object density $p_{\kappa,h}$ is calculated. The final recognition result $(\widehat{\kappa}, \widehat{\boldsymbol{\phi}}, \widehat{\boldsymbol{t}})$ corresponds to the highest density normalized by the quality measure $Q(p_{\kappa,h})$.

are created in the training phase (Chapter 4). In $\mathcal{M}_{\kappa,-1}$ the objects are represented by feature vectors calculated for local neighborhoods of size 2×2 pixels, $\mathcal{M}_{\kappa,-2}$ uses neighborhoods of size 4×4 pixels, and $\mathcal{M}_{\kappa,-3}$ computes the features from neighborhoods of size 8×8 pixels. Therefore, in general, a statistical object model $\mathcal{M}_{\kappa,\widehat{s}}$ describes objects by feature vectors extracted for local neighborhoods of size $2^{|\widehat{s}|}\times 2^{|\widehat{s}|}$ pixels.

The classification and localization result (5.16) determined on the resolution level $L_{\widehat{s}}$ is denoted by $(\widehat{\kappa}_{\widehat{s}}, \widehat{\boldsymbol{\phi}}_{\widehat{s}}, \widehat{\boldsymbol{t}}_{\widehat{s}})$. The recognition algorithm presented in [Grz05b] takes into consideration all statistical models $\mathcal{M}_{\kappa,\widehat{s}=-1,-2,-3}$ trained for each object class Ω_{κ}. It consists of the following steps:

1. *Recognition on* $L_{\widehat{s}} = L_{-3}$
   First, the class index $\widehat{\kappa}_{-3}$ and the pose $(\widehat{\boldsymbol{\phi}}_{-3}, \widehat{\boldsymbol{t}}_{-3})$ of the object in the test image $\boldsymbol{f}$ are determined with the maximum likelihood (ML) estimation (5.16) for the resolution level L_{-3}, i. e., using the statistical object models $\mathcal{M}_{\kappa,-3}$.

2. *Search Space for* $L_{\widehat{s}} = L_{-2}$
   Second, the search space for the resolution level L_{-2} is determined based on the localization result $(\widehat{\boldsymbol{\phi}}_{-3}, \widehat{\boldsymbol{t}}_{-3})$ from the resolution L_{-3}. The range for rotation parameters $\boldsymbol{\phi} = (\phi_x, \phi_y, \phi_z)^{\mathrm{T}}$ is defined by $\pm 10°$, the range for internal translations $\boldsymbol{t}_{\mathrm{int}} = (t_x, t_y)^{\mathrm{T}}$ by ± 4 pixels, and the range for scaling t_z is determined with $\pm 10\%$.

3. *Recognition on* $L_{\widehat{s}} = L_{-2}$
   Third, the recognition result $(\widehat{\kappa}_{-2}, \widehat{\boldsymbol{\phi}}_{-2}, \widehat{\boldsymbol{t}}_{-2})$ for the resolution level L_{-2} is found using the statistical object models $\mathcal{M}_{\kappa,-2}$. The maximization (5.16) is performed for all object classes $\Omega_{\kappa=1,\ldots,N_{\Omega}}$ and the search space defined in the previous point.

4. *Search Space for* $L_{\widehat{s}} = L_{-1}$
   Fourth, the search space for the resolution level L_{-1} is determined based on the localization result $(\widehat{\boldsymbol{\phi}}_{-2}, \widehat{\boldsymbol{t}}_{-2})$ from the resolution L_{-2}. The range for rotation parameters $\boldsymbol{\phi} = (\phi_x, \phi_y, \phi_z)^{\mathrm{T}}$ is defined by $\pm 5°$, the range for internal translations $\boldsymbol{t}_{\mathrm{int}} = (t_x, t_y)^{\mathrm{T}}$ by ± 2 pixels, and the range for scaling t_z is determined with $\pm 5\%$.

5. *Recognition on* $L_{\widehat{s}} = L_{-1}$
   The final classification and localization result $(\widehat{\kappa}_{-1}, \widehat{\boldsymbol{\phi}}_{-1}, \widehat{\boldsymbol{t}}_{-1})$ is obtained for the resolution level L_{-1} using statistical object models $\mathcal{M}_{\kappa,-1}$. The maximization (5.16) is performed for all object classes $\Omega_{\kappa=1,\ldots,N_{\Omega}}$ and the search space defined in the previous point.

Using the combination of object models for different wavelet resolutions, the execution time of the recognition algorithm increases, but significantly better classification and localization rates are obtained [Grz05b].

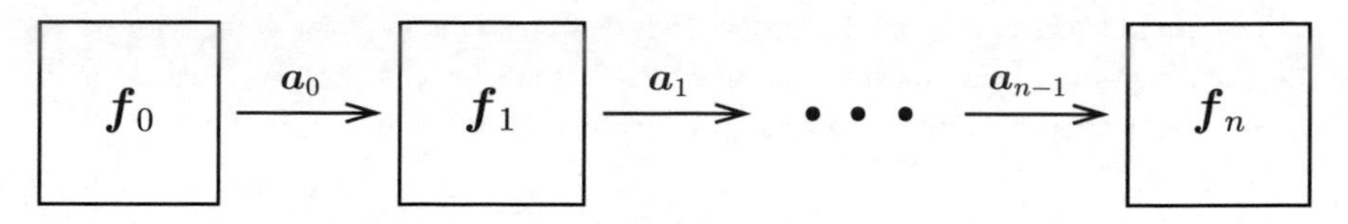

Figure 5.5: Sequence of different object views $(\boldsymbol{f}_0, \boldsymbol{f}_1, \ldots, \boldsymbol{f}_n)$ and the camera movements $(\boldsymbol{a}_0, \boldsymbol{a}_1, \ldots, \boldsymbol{a}_{n-1})$ between them.

5.3 Fusion of Multiple Views

Section 5.2 presents the recognition algorithm dealing with single-object scenes and using one image $\boldsymbol{f}$ for object classification and localization. However, as can be seen in Figure 1.4 (Chapter 1), some objects cannot be distinguished using only one point of view. In this case a fusion of multiple views is required in order to classify and localize objects [Dei05]. Therefore, in this section still exactly one resulting object class $\Omega_{\widehat{\kappa}}$ is expected, but a fusion of multiple different views

$$\langle \boldsymbol{f} \rangle_n = (\boldsymbol{f}_0, \boldsymbol{f}_1, \ldots, \boldsymbol{f}_n) \tag{5.19}$$

of the object is considered for recognition. The camera movements between the viewpoints are denoted by

$$\langle \boldsymbol{a} \rangle_{n-1} = (\boldsymbol{a}_0, \boldsymbol{a}_1, \ldots, \boldsymbol{a}_{n-1}) \tag{5.20}$$

and represent the object poses relative to each other (see Figure 5.5). The object class Ω_κ and pose $(\boldsymbol{\phi}, \boldsymbol{t})$ in the image $\boldsymbol{f}_n$ is called *object state* in this section and is denoted by

$$\boldsymbol{q}_n = (\Omega_\kappa, \phi_x, \phi_y, \phi_z, t_x, t_y, t_z)^{\mathrm{T}} \tag{5.21}$$

The object state $\boldsymbol{q}_n$ in the image $\boldsymbol{f}_n$ contains the discrete information about the object class (Ω_κ) and the continuous information about the object pose ($\boldsymbol{\phi} = (\phi_x, \phi_y, \phi_z)^{\mathrm{T}}$ and $\boldsymbol{t} = (t_x, t_y, t_z)^{\mathrm{T}}$). The task of object recognition with fusion of multiple views can be solved by the object state $\boldsymbol{q}_n$ estimation for the given image sequence $\langle \boldsymbol{f} \rangle_n$ and camera movements $\langle \boldsymbol{a} \rangle_{n-1}$. This can be done by maximization of the following density function

$$p'_{\boldsymbol{q}_n} = p(\boldsymbol{q}_n | \boldsymbol{f}_n, \boldsymbol{a}_{n-1}, \ldots, \boldsymbol{a}_0, \boldsymbol{f}_0) \tag{5.22}$$

over all possible object states $\boldsymbol{q}_n$. In Section 5.3.1 the recursive propagation of the density function defined by (5.22) is explained. Section 5.3.2 describes the application of *condensation algorithm* for object recognition with fusion of multiple views.

5.3.1 Recursive Density Propagation

The object state density (5.22) can be modified to

$$p'_{\boldsymbol{q}_n} = \frac{1}{\tau_n} p(\boldsymbol{q}_n | \boldsymbol{a}_{n-1}, \boldsymbol{f}_{n-1} \ldots, \boldsymbol{a}_0, \boldsymbol{f}_0) p(\boldsymbol{f}_n | \boldsymbol{q}_n) \quad , \tag{5.23}$$

where

$$\tau_n = p(\boldsymbol{f}_n, \boldsymbol{a}_{n-1}, \ldots, \boldsymbol{a}_0, \boldsymbol{f}_0) \tag{5.24}$$

denotes a normalizing constant that is left out in the following considerations. Instead of $p'_{\boldsymbol{q}_n}$ the term $p_{\boldsymbol{q}_n} = \tau_n p'_{\boldsymbol{q}_n}$ is taken into account for the recursive density propagation[2]

$$p_{\boldsymbol{q}_n} = \tau_n p'_{\boldsymbol{q}_n} = \underbrace{p(\boldsymbol{q}_n | \boldsymbol{a}_{n-1}, \boldsymbol{f}_{n-1} \ldots, \boldsymbol{a}_0, \boldsymbol{f}_0)}_{(\star)} p(\boldsymbol{f}_n | \boldsymbol{q}_n) \quad . \tag{5.25}$$

Since the informations about the object views $(\boldsymbol{f}_0, \boldsymbol{f}_1, \ldots, \boldsymbol{f}_{n-1})$ and the camera movements $(\boldsymbol{a}_0, \boldsymbol{a}_1, \ldots, \boldsymbol{a}_{n-2})$ are stored in the object state $\boldsymbol{q}_{n-1}$, the following simplification can be done

$$p(\boldsymbol{q}_n | \boldsymbol{q}_{n-1}, \boldsymbol{f}_{n-1}, \boldsymbol{a}_{n-2}, \ldots, \boldsymbol{a}_0, \boldsymbol{f}_0) = p(\boldsymbol{q}_n | \boldsymbol{q}_{n-1}, \boldsymbol{a}_{n-1}) \quad . \tag{5.26}$$

Under this so-called Markov assumption for the state transition, the term $(\star)$ within (5.25) can be recursively rewritten as

$$p(\boldsymbol{q}_n | \boldsymbol{a}_{n-1}, \boldsymbol{f}_{n-1} \ldots, \boldsymbol{a}_0, \boldsymbol{f}_0) = \int p(\boldsymbol{q}_n | \boldsymbol{q}_{n-1}, \boldsymbol{a}_{n-1}) \underbrace{p(\boldsymbol{q}_{n-1} | \boldsymbol{f}_{n-1}, \boldsymbol{a}_{n-2}, \ldots, \boldsymbol{a}_0, \boldsymbol{f}_0)}_{(\star\star)} d\boldsymbol{q}_{n-1} \quad . \tag{5.27}$$

Considering (5.25), the term $(\star\star)$ in (5.27) can be replaced by $p_{\boldsymbol{q}_{n-1}}$ and the object state density (5.25) modifies to

$$p_{\boldsymbol{q}_n} = p(\boldsymbol{f}_n | \boldsymbol{q}_n) \int p(\boldsymbol{q}_n | \boldsymbol{q}_{n-1}, \boldsymbol{a}_{n-1}) p_{\boldsymbol{q}_{n-1}} d\boldsymbol{q}_{n-1} \quad . \tag{5.28}$$

[2] Although $p_{\boldsymbol{q}_n}$ differs from $p'_{\boldsymbol{q}_n}$ about the constant factor τ_n, it is also called object state density.

(5.28) is called recursive density propagation. The recursivity lies in the fact that the probability $p_{\boldsymbol{q}_n}$ for the object state $\boldsymbol{q}_n$ in the image $\boldsymbol{f}_n$ is expressed using the object state probability $p_{\boldsymbol{q}_{n-1}}$. In reality only a finite number of object pose hypotheses $(\boldsymbol{\phi}_h, \boldsymbol{t}_h)$ can be evaluated in the recognition phase. Therefore, the object state can be considered as a vector containing only discrete elements and denoted by

$$\boldsymbol{q}_{n,h} = (\Omega_{\kappa,h}, \phi_{x,h}, \phi_{y,h}, \phi_{z,h}, t_{x,h}, t_{y,h}, t_{z,h})^{\mathrm{T}} \quad . \tag{5.29}$$

The recursive density propagation (5.28) takes in this case its discrete form

$$p_{n,h} = p(\boldsymbol{f}_n|\boldsymbol{q}_{n,h}) \sum_{i=1}^{N_h} \{p(\boldsymbol{q}_{n,h}|\boldsymbol{q}_{n-1,i}, \boldsymbol{a}_{n-1}) p_{n-1,h}\} \quad , \tag{5.30}$$

where N_h is the number of all object state hypotheses $\boldsymbol{q}_{n,h=1,\ldots,N_h}$. The object classification and localization in the image $\boldsymbol{f}_n$ using the fusion of multiple views can be solved by maximization of the object state probability over all object states for this image, which can be written in the continuous form using (5.28)

$$\widehat{\boldsymbol{q}}_n = \underset{\boldsymbol{q}_n}{\operatorname{argmax}} \left\{ p(\boldsymbol{f}_n|\boldsymbol{q}_n) \int p(\boldsymbol{q}_n|\boldsymbol{q}_{n-1}, \boldsymbol{a}_{n-1}) p_{\boldsymbol{q}_{n-1}} d\boldsymbol{q}_{n-1} \right\} \tag{5.31}$$

or in the discrete form using (5.30)

$$\boldsymbol{q}_{n,\widehat{h}} \quad \Leftarrow \quad \widehat{h} = \underset{h}{\operatorname{argmax}} \left\{ p(\boldsymbol{f}_n|\boldsymbol{q}_{n,h}) \sum_{i=1}^{N_h} \{p(\boldsymbol{q}_{n,h}|\boldsymbol{q}_{n-1,i}, \boldsymbol{a}_{n-1}) p_{n-1,h}\} \right\} \quad . \tag{5.32}$$

The evaluation of the probability $p(\boldsymbol{f}_n|\boldsymbol{q}_{n,h})$ for the single test image $\boldsymbol{f}_n$ given the object state hypothesis $\boldsymbol{q}_{n,h}$ is described in detail in Section 5.2.1 and finally given with (5.10), whereas

$$p(\boldsymbol{f}_n|\boldsymbol{q}_{n,h}) = p_n(C_{O_\kappa}|\widehat{\boldsymbol{B}}_\kappa, \boldsymbol{\phi}_h, \boldsymbol{t}_h) \quad . \tag{5.33}$$

The test image $\boldsymbol{f}_n$ is represented by the set of object feature vectors C_{O_κ}, while the object state hypothesis $\boldsymbol{q}_{n,h}$ is given by the class hypothesis $\widehat{\boldsymbol{B}}_\kappa$ and the pose hypothesis $(\boldsymbol{\phi}_h, \boldsymbol{t}_h)$. The index n of the density p corresponds to the test image number. The recursive estimation of the density values $p_{n,h}$ (5.30) is performed with the condensation algorithm in this work, which is explained in detail in the next section.

5.3.2 Condensation Algorithm

After the presentation of the density propagation theory in the previous section, the practical realization of the fusion of multiple views follows in this section. The classic approach for solving the recursive density propagation (5.30) is based on the *Kalman filter* [Kal60, BS88]. However, the necessary assumption for the Kalman filter, i. e., that $p(\boldsymbol{f}_n|\boldsymbol{q}_{n,h})$ is normally distributed, is often not valid in real applications of computer vision due to object ambiguities, sensor noise, occlusion, etc. In a real world environment this distribution takes usually a very complex form and is analytically not computable [Grz03]. Therefore, in the present work the so-called condensation algorithm [Isa98] is used to estimate the recursive object state density (5.30).

Let's define a *particle set* Z_n for the test image $\boldsymbol{f}_n$ by

$$Z_n = \{\langle \boldsymbol{q}_{n,1}, p_{n,1}\rangle, \ldots, \langle \boldsymbol{q}_{n,h}, p_{n,h}\rangle, \ldots, \langle \boldsymbol{q}_{n,N_h}, p_{n,N_h}\rangle\} \quad , \tag{5.34}$$

where N_h is the number of all object state hypotheses[3] $\boldsymbol{q}_{n,h}$, while $\langle \boldsymbol{q}_{n,h}, p_{n,h}\rangle$ represents a single particle containing the object state $\boldsymbol{q}_{n,h}$ and its probability $p_{n,h}$ (5.30) for the test image $\boldsymbol{f}_n$. In the practical realization of the condensation algorithm one starts with the first test image $\boldsymbol{f}_0$ of the sequence $\langle \boldsymbol{f}\rangle_n$ and determines the particle sets for the remaining images recursively. Three beginning steps of the algorithm are described below:

1. *Particle Set Z_0 for Image $\boldsymbol{f}_0$*
According to (5.34) the initial particle set Z_0 for the first test image $\boldsymbol{f}_0$ in the sequence can be written as

$$Z_0 = \{\langle \boldsymbol{q}_{0,1}, p_{0,1}\rangle, \ldots, \langle \boldsymbol{q}_{0,h}, p_{0,h}\rangle, \ldots, \langle \boldsymbol{q}_{0,N_h}, p_{0,N_h}\rangle\} \quad . \tag{5.35}$$

The initial probabilities $p_{0,h}$ for the object state hypotheses $\boldsymbol{q}_{0,h}$ have to be estimated using only one image $\boldsymbol{f}_0$. Therefore, the equation (5.30) degrades in this case to

$$p_{0,h} = p(\boldsymbol{f}_0|\boldsymbol{q}_{0,h}) = p_0(C_{O_\kappa}|\widehat{\boldsymbol{B}}_\kappa, \boldsymbol{\phi}_h, \boldsymbol{t}_h) \quad . \tag{5.36}$$

2. *Particle Set Z_1 for Image $\boldsymbol{f}_1$*
The particle set Z_1 for the second test image $\boldsymbol{f}_1$ looks as follows

$$Z_1 = \{\langle \boldsymbol{q}_{1,1}, p_{1,1}\rangle, \ldots, \langle \boldsymbol{q}_{1,h}, p_{1,h}\rangle, \ldots, \langle \boldsymbol{q}_{1,N_h}, p_{1,N_h}\rangle\} \quad , \tag{5.37}$$

[3] Considering only the classification task, N_h is just the number of object classes N_Ω ($N_h = N_\Omega$).

whereas the probabilities $p_{1,h}$ (5.30) depend in this case on the probabilities $p_{0,h}$ in the first step of the algorithm

$$p_{1,h} = p(\boldsymbol{f}_1|\boldsymbol{q}_{1,h}) \sum_{i=1}^{N_h} \{p(\boldsymbol{q}_{1,h}|\boldsymbol{q}_{0,i}, \boldsymbol{a}_0)p_{0,h}\} \quad . \tag{5.38}$$

3. *Particle Set Z_2 for Image $\boldsymbol{f}_2$*
   The particle set Z_2 for the third test image $\boldsymbol{f}_2$ can be denoted by

$$Z_2 = \{\langle \boldsymbol{q}_{2,1}, p_{2,1}\rangle, \ldots, \langle \boldsymbol{q}_{2,h}, p_{2,h}\rangle, \ldots, \langle \boldsymbol{q}_{2,N_h}, p_{2,N_h}\rangle\} \quad , \tag{5.39}$$

   The weights $p_{2,h}$ are in this step calculated with

$$p_{2,h} = p(\boldsymbol{f}_2|\boldsymbol{q}_{2,h}) \sum_{i=1}^{N_h} \{p(\boldsymbol{q}_{2,h}|\boldsymbol{q}_{1,i}, \boldsymbol{a}_1)p_{1,h}\} \quad . \tag{5.40}$$

Repeating the steps one by one, the probability distribution $p_{n,h}$ for all object state hypotheses $\boldsymbol{q}_{n,h}$ in any test image $\boldsymbol{f}_n$ can be determined and represented by the particle set Z_n (5.34). The resulting object state $\boldsymbol{q}_{n,\widehat{h}}$ (classification $\widehat{\kappa}$ and localization $(\widehat{\boldsymbol{\phi}}, \widehat{\boldsymbol{t}})$ result in image $\boldsymbol{f}_n$) corresponds to the highest density value $p_{n,\widehat{h}}$ (5.32)

$$\boldsymbol{q}_{n,\widehat{h}} \quad \Leftarrow \quad \langle \boldsymbol{q}_{n,\widehat{h}}, \max\{p_{n,1}, \ldots, p_{n,h}, \ldots, p_{n,N_h}\}\rangle \quad . \tag{5.41}$$

5.4 Multi-Object Scenes

In the two previous sections algorithms for object recognition in single-object scenes were presented; while the following two sections discuss the classification and localization problem for multi-object scenes. The present section assumes the uniform distribution of the a-priori occurrence probabilities for all object classes

$$p(\Omega_1) = \cdots = p(\Omega_\kappa) = \cdots = p(\Omega_{N_\Omega}) \tag{5.42}$$

and does not take into consideration the context models learned in Section 4.4. Moreover, in contrast to Section 5.3, only one test scene is regarded for determining the object classes and poses. In the recognition task for multi-object scenes, not only the classes of objects and their poses have to be determined. Since the number of objects in a scene is a-priori unknown, it also

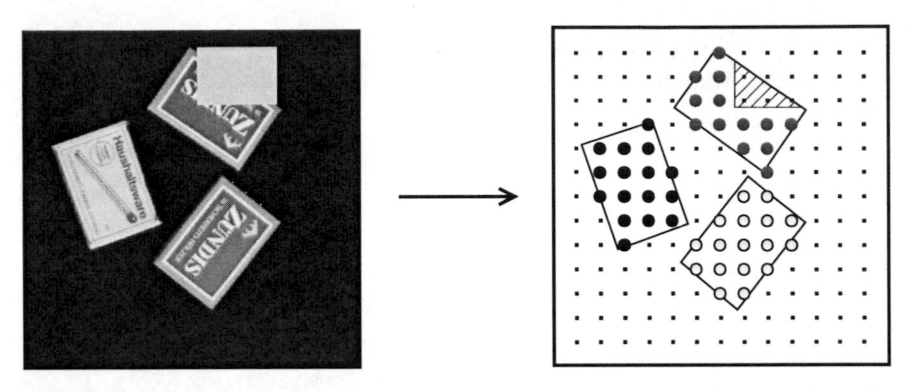

Figure 5.6: In the recognition algorithm for multi-object scenes presented in [Grz04a] a global assignment function assigns the feature vectors to the objects, whereas one feature vector corresponds maximal to one object in a scene. After a feature vector was assigned to an object, it is not taken into account in the next steps of the serial search algorithm.

must be estimated. An example of a real multi-object scene with heterogeneous background can be found in Figure 1.1 in Chapter 1.

In [Grz04a] an object classification and localization approach for multi-object scenes based on the so-called *serial search algorithm* is presented. The authors represent objects by 2D local feature vectors computed with the wavelet transformation using gray level images, i. e., similarly to Section 4.2. In order to solve the recognition problem for multi-object scenes, the so-called *global assignment function* is introduced. This function assigns feature vectors extracted in the test image to the objects occurring in it, which is depicted in Figure 5.6. One feature vector is assigned to at most one object in a scene. Once a feature vector has already been assigned, it is labeled and not used in the next steps of the serial search algorithm. First, the algorithm estimates the optimal poses $({}^1\widehat{\boldsymbol{\phi}}_\kappa, {}^1\widehat{\boldsymbol{t}}_\kappa)$ for all possible object classes Ω_κ with the maximization algorithm illustrated in Figure 5.4. The prefix "1" denotes searching for the first object in the scene. Subsequently, the system starts to search for the second object in the scene, whereas the feature vectors labeled by the global assignment function in the first step are not taken into consideration. The serial search algorithm is terminated if there are no more valid class Ω_κ and pose $(\boldsymbol{\phi}_h, \boldsymbol{t}_h)$ hypotheses. The validity of a class Ω_κ and pose $(\boldsymbol{\phi}_h, \boldsymbol{t}_h)$ hypothesis is defined as follows

$$\frac{N_\kappa(\boldsymbol{\phi}_h, \boldsymbol{t}_h) - N'_\kappa(\boldsymbol{\phi}_h, \boldsymbol{t}_h)}{N_\kappa(\boldsymbol{\phi}_h, \boldsymbol{t}_h)} \quad \begin{cases} \geq S_v & \Rightarrow \text{ hypothesis valid} \\ < S_v & \Rightarrow \text{ hypothesis not valid} \end{cases} . \tag{5.43}$$

$N_\kappa(\boldsymbol{\phi}_h, \boldsymbol{t}_h)$ (5.14) is the number of all object feature vectors $\boldsymbol{c}_m \in C_{O_\kappa}$ for the object class Ω_κ and the object pose $(\boldsymbol{\phi}_h, \boldsymbol{t}_h)$, while $N'_\kappa(\boldsymbol{\phi}_h, \boldsymbol{t}_h)$ denotes the number of object feature vectors that, according to revision (5.9), do not really describe the object ($\boldsymbol{c}_m \notin C'_{O_\kappa}$). In other words, $N_\kappa(\boldsymbol{\phi}_h, \boldsymbol{t}_h)$ is the number of elements of the set C_{O_κ}, while $N'_\kappa(\boldsymbol{\phi}_h, \boldsymbol{t}_h)$ is the number of elements of the set $C_{O_\kappa} \setminus C'_{O_\kappa}$ for the object class Ω_κ and the object pose $(\boldsymbol{\phi}_h, \boldsymbol{t}_h)$. The optimal threshold S_v is determined in an experimental way [Grz04a]. On the one hand, the method introduced in [Grz04a] is very time-consuming, but, on the other hand, it allows multiple occurrence of the same object in the scene (see Figure 5.6).

The serial search algorithm for multi-object scenes [Grz04a] is simplified in the scope of this work. The new approach yields comparable results to [Grz04a] (Section 6.5), but it is significantly faster. Regarding (5.42) and the assumption that all object pose hypotheses $(\boldsymbol{\phi}_h, \boldsymbol{t}_h)$ are also equiprobable, the maximum a-posteriori estimation (MAP) can be modified to the maximum likelihood estimation (ML), which is explained in detail in Section 3.1.5 (3.24) and mentioned also in Section 5.2.2 (5.13). Figure 5.4 illustrates the maximum likelihood estimation for object classification and localization. First, the ML algorithm estimates the optimal pose parameters $(\widehat{\boldsymbol{\phi}}_\kappa, \widehat{\boldsymbol{t}}_\kappa)$ for all object classes Ω_κ considered in the recognition task by maximizing the normalized object density value according to (5.16). This can be expressed with the following maximization terms

$$\begin{aligned}
(\widehat{\boldsymbol{\phi}}_1, \widehat{\boldsymbol{t}}_1) &= \underset{(\boldsymbol{\phi}_h, \boldsymbol{t}_h)}{\operatorname{argmax}} \, Q(p(C_{O_1} | \widehat{\boldsymbol{B}}_1, \boldsymbol{\phi}_h, \boldsymbol{t}_h)) \\
&\dots \\
(\widehat{\boldsymbol{\phi}}_\kappa, \widehat{\boldsymbol{t}}_\kappa) &= \underset{(\boldsymbol{\phi}_h, \boldsymbol{t}_h)}{\operatorname{argmax}} \, Q(p(C_{O_\kappa} | \widehat{\boldsymbol{B}}_\kappa, \boldsymbol{\phi}_h, \boldsymbol{t}_h)) \quad . \\
&\dots \\
(\widehat{\boldsymbol{\phi}}_{N_\Omega}, \widehat{\boldsymbol{t}}_{N_\Omega}) &= \underset{(\boldsymbol{\phi}_h, \boldsymbol{t}_h)}{\operatorname{argmax}} \, Q(p(C_{O_{N_\Omega}} | \widehat{\boldsymbol{B}}_{N_\Omega}, \boldsymbol{\phi}_h, \boldsymbol{t}_h))
\end{aligned} \tag{5.44}$$

The normalized object density values for the optimal pose parameters can be written in short forms as follows

$$\begin{aligned}
\widehat{Q}_1 &= Q(p(C_{O_1} | \widehat{\boldsymbol{B}}_1, \widehat{\boldsymbol{\phi}}_1, \widehat{\boldsymbol{t}}_1)) \\
&\dots \\
\widehat{Q}_\kappa &= Q(p(C_{O_\kappa} | \widehat{\boldsymbol{B}}_\kappa, \widehat{\boldsymbol{\phi}}_\kappa, \widehat{\boldsymbol{t}}_\kappa)) \quad . \\
&\dots \\
\widehat{Q}_{N_\Omega} &= Q(p(C_{O_{N_\Omega}} | \widehat{\boldsymbol{B}}_{N_\Omega}, \widehat{\boldsymbol{\phi}}_{N_\Omega}, \widehat{\boldsymbol{t}}_{N_\Omega}))
\end{aligned} \tag{5.45}$$

These normalized object densities $\widehat{Q}_{\kappa=1,\ldots,N_\Omega}$ are now sorted from the highest to the lowest value, i. e., in a non-increasing way

$$\underbrace{\widehat{Q}_{\kappa_1} \geq \widehat{Q}_{\kappa_2}}_{d_1} \geq \ldots \geq \underbrace{\widehat{Q}_{\kappa_i} \geq \widehat{Q}_{\kappa_{i+1}}}_{d_i} \geq \ldots \geq \widehat{Q}_{\kappa_I} \quad , \tag{5.46}$$

where $I = N_\Omega$ and d_i is a difference between neighboring elements

$$d_i = d(\widehat{Q}_{\kappa_i}, \widehat{Q}_{\kappa_{i+1}}) = \widehat{Q}_{\kappa_i} - \widehat{Q}_{\kappa_{i+1}} \quad . \tag{5.47}$$

Finally, the index $\widehat{i}$ of the highest distance $d_{\widehat{i}}$ ($\forall\, i \neq \widehat{i}\, :\, d_i \leq d_{\widehat{i}}$) can be easily estimated with the following formula

$$\widehat{i} = \operatorname*{argmax}_i d_i \tag{5.48}$$

and it is interpreted as the number of objects occurring in the multi-object scene $\boldsymbol{f}$. Hence, the final recognition result in the multi-object scene $\boldsymbol{f}$ are the following object classes and poses

$$\begin{array}{ll} \text{first object} & (\kappa_1, \widehat{\boldsymbol{\phi}}_{\kappa_1}, \widehat{\boldsymbol{t}}_{\kappa_1}) \\ \text{second object} & (\kappa_2, \widehat{\boldsymbol{\phi}}_{\kappa_2}, \widehat{\boldsymbol{t}}_{\kappa_2}) \\ \vdots & \\ \text{last object} & (\kappa_{\widehat{i}}, \widehat{\boldsymbol{\phi}}_{\kappa_{\widehat{i}}}, \widehat{\boldsymbol{t}}_{\kappa_{\widehat{i}}}) \end{array} \quad . \tag{5.49}$$

In order to evaluate the recognition algorithm for multi-object scenes, not only the object classification result Ω_{κ_i} and the object localization result $(\widehat{\boldsymbol{\phi}}_{\kappa_i}, \widehat{\boldsymbol{t}}_{\kappa_i})$ have to be verified. The number $\widehat{i}$ of objects found in the scene $\boldsymbol{f}$ must also be checked (Section 6.5).

5.5 Scenes with Context Dependencies

The present section deals also with multi-object scenes, but, in contrast to Section 5.4, it takes into consideration context dependencies between objects, which were statistically modeled in Section 4.4. Having additional knowledge about the environment, in which an image was taken, the occurrence of some objects might be more likely than the occurrence of the others. Three example contexts (office, kitchen, and nursery) are depicted in Figure 4.8. In the kitchen context objects like plates, knifes, or forks occur more likely than, e. g., punchers, staplers, or pens, which are rather found in an office. However, in the recognition phase there is no a-priori knowledge about the context Υ_ι (4.74), in which the test image $\boldsymbol{f}$ was taken. Hence, the system does not a-priori know, which of the context models $\mathcal{M}_{\iota=1,\dots,N_\Upsilon}$ learned in the training phase (Section 4.4) should be applied for object recognition in a particular test scene $\boldsymbol{f}$. For this reason the algorithm for multi-object scenes with context dependencies is split into two phases. In the first phase the algorithm determines automatically the context $\Upsilon_{\widehat{\iota}}$ for the image $\boldsymbol{f}$. In the second phase the approach classifies and localizes objects in the image $\boldsymbol{f}$ using the statistical context model $\mathcal{M}_{\widehat{\iota}}$ learned for this context $\Upsilon_{\widehat{\iota}}$ in the training phase (Section 4.4).

Searching for the first object Ω_{κ_1} in the multi-object scene $\boldsymbol{f}$, the algorithm does not use any context information and, similarly to previous sections in this chapter, it assumes equal a-priori probabilities

$$p(\Omega_1) = \cdots = p(\Omega_\kappa) = \cdots = p(\Omega_{N_\Omega}) \tag{5.50}$$

for all object classes $\Omega_{\kappa=1,\dots,N_\Omega}$ considered in the recognition task. The class κ_1 and the pose $(\widehat{\boldsymbol{\phi}}_1, \widehat{\boldsymbol{t}}_1)$ of the first object in the image $\boldsymbol{f}$ is determined by maximization of the normalized object density value with (5.16), which is illustrated in Figure 5.4. As mentioned in Section 1.1, it is assumed that at least one of the objects from the set $\Omega = \{\Omega_1, \Omega_2, \dots, \Omega_\kappa, \dots, \Omega_{N_\Omega}\}$ occurs in the image $\boldsymbol{f}$, i. e., $\widehat{i} \geq 1$ (5.48). Subsequently, the context $\Upsilon_{\widehat{\iota}}$ for the scene $\boldsymbol{f}$ (or just the context number $\widehat{\iota}$) is determined using the statistical context models $\mathcal{M}_{\iota=1,\dots,N_\Upsilon}$ learned as shown in Section 4.4. Each context model $\mathcal{M}_\iota$ contains the trained a-priori probabilities $p_\iota(\Omega_\kappa)$ for all object classes $\Omega_{\kappa=1,\dots,N_\Omega}$, whereas

$$p_\iota(\Omega_1) \neq \cdots \neq p_\iota(\Omega_\kappa) \neq \cdots \neq p_\iota(\Omega_{N_\Omega}) \quad . \tag{5.51}$$

Therefore, using the context models $\mathcal{M}_{\iota=1,\dots,N_\Upsilon}$ it is possible to determine the a-priori density $p_{\iota=1,\dots,N_\Upsilon}(\Omega_{\kappa_1})$ of the first object class Ω_{κ_1} for all contexts $\Upsilon_{\iota=1,\dots,N_\Upsilon}$. The highest value of these densities $p_{\iota=1,\dots,N_\Upsilon}(\Omega_{\kappa_1})$ decides about the context $\Upsilon_{\widehat{\iota}}$ of the multi-object scene $\boldsymbol{f}$

$$\widehat{\iota} = \underset{\iota}{\operatorname{argmax}}\, p_\iota(\Omega_{\kappa_1}) \quad . \tag{5.52}$$

Looking for further objects $(\Omega_{\kappa_2}, \Omega_{\kappa_3}, \dots, \Omega_{\kappa_{\widehat{i}}})$ in the image $\boldsymbol{f}$ the statistical context model $\mathcal{M}_{\widehat{\iota}}$ learned for the context $\Upsilon_{\widehat{\iota}}$ is used and the following a-priori probabilities for object occurrence

$$p_{\widehat{\iota}}(\Omega_1) \neq \cdots \neq p_{\widehat{\iota}}(\Omega_\kappa) \neq \cdots \neq p_{\widehat{\iota}}(\Omega_{N_\Omega}) \tag{5.53}$$

are taken into consideration.

After the context $\Upsilon_{\widehat{i}}$ for the scene $\boldsymbol{f}$ has already been determined, the system estimates the optimal pose parameters $(\widehat{\boldsymbol{\phi}}_\kappa, \widehat{\boldsymbol{t}}_\kappa)$ for all object classes Ω_κ considered in the recognition task similarly to Section 5.4

$$\begin{aligned}
(\widehat{\boldsymbol{\phi}}_1, \widehat{\boldsymbol{t}}_1) &= \underset{(\boldsymbol{\phi}_h, \boldsymbol{t}_h)}{\text{argmax}}\, Q(p(C_{O_1}|\widehat{\boldsymbol{B}}_1, \boldsymbol{\phi}_h, \boldsymbol{t}_h)) \\
&\cdots \\
(\widehat{\boldsymbol{\phi}}_\kappa, \widehat{\boldsymbol{t}}_\kappa) &= \underset{(\boldsymbol{\phi}_h, \boldsymbol{t}_h)}{\text{argmax}}\, Q(p(C_{O_\kappa}|\widehat{\boldsymbol{B}}_\kappa, \boldsymbol{\phi}_h, \boldsymbol{t}_h)) \quad . \\
&\cdots \\
(\widehat{\boldsymbol{\phi}}_{N_\Omega}, \widehat{\boldsymbol{t}}_{N_\Omega}) &= \underset{(\boldsymbol{\phi}_h, \boldsymbol{t}_h)}{\text{argmax}}\, Q(p(C_{O_{N_\Omega}}|\widehat{\boldsymbol{B}}_{N_\Omega}, \boldsymbol{\phi}_h, \boldsymbol{t}_h))
\end{aligned} \tag{5.54}$$

The use of the maximum likelihood estimation (ML) (5.12) in (5.54) is allowed, because for the given object class Ω_κ all object pose hypotheses $(\boldsymbol{\phi}_h, \boldsymbol{t}_h)$ are assumed to be a-priori equiprobable. However, due to the non-uniform distribution of the object occurrence densities $p_{\widehat{\iota}}(\Omega_\kappa)$ (5.53) in the context $\Upsilon_{\widehat{\iota}}$ the normalized object density values for the optimal pose parameters cannot be written like in (5.45). They have to be weighted with the a-priori probabilities $p_{\widehat{\iota}}(\Omega_\kappa)$ learned for the context $\Upsilon_{\widehat{\iota}}$ in the training phase (Section 4.4) and stored in the statistical object model $\mathcal{M}_{\widehat{\iota}}$

$$\begin{aligned}
\widehat{Q}_{\widehat{\iota},1} &= Q\{p_{\widehat{\iota}}(\Omega_1)p(C_{O_1}|\widehat{\boldsymbol{B}}_1, \widehat{\boldsymbol{\phi}}_1, \widehat{\boldsymbol{t}}_1)\} \\
&\cdots \\
\widehat{Q}_{\widehat{\iota},\kappa} &= Q\{p_{\widehat{\iota}}(\Omega_\kappa)p(C_{O_\kappa}|\widehat{\boldsymbol{B}}_\kappa, \widehat{\boldsymbol{\phi}}_\kappa, \widehat{\boldsymbol{t}}_\kappa)\} \quad . \\
&\cdots \\
\widehat{Q}_{\widehat{\iota},N_\Omega} &= Q\{p_{\widehat{\iota}}(\Omega_{N_\Omega})p(C_{O_{N_\Omega}}|\widehat{\boldsymbol{B}}_{N_\Omega}, \widehat{\boldsymbol{\phi}}_{N_\Omega}, \widehat{\boldsymbol{t}}_{N_\Omega})\}
\end{aligned} \tag{5.55}$$

These normalized and weighted object densities $\widehat{Q}_{\widehat{\iota},\kappa=1,\ldots,N_\Omega}$ are now sorted from the highest to the lowest value, i. e., in a non-increasing way

$$\underbrace{\widehat{Q}_{\kappa_1} \geq \widehat{Q}_{\kappa_2}}_{d_1} \geq \ldots \geq \underbrace{\widehat{Q}_{\kappa_i} \geq \widehat{Q}_{\kappa_{i+1}}}_{d_i} \geq \ldots \geq \widehat{Q}_{\kappa_I} \quad , \tag{5.56}$$

where $I = N_\Omega$ and d_i is a difference between neighboring elements

$$d_i = d(\widehat{Q}_{\kappa_i}, \widehat{Q}_{\kappa_{i+1}}) = \widehat{Q}_{\kappa_i} - \widehat{Q}_{\kappa_{i+1}} \quad . \tag{5.57}$$

The index $\widehat{i}$ of the highest distance $d_{\widehat{i}}$ $(\forall\, i \neq \widehat{i} \,:\, d_i \leq d_{\widehat{i}})$ is interpreted as the number of objects found in the multi-object scene $\boldsymbol{f}$ and is calculated by

$$\widehat{i} = \underset{i}{\operatorname{argmax}}\, d_i \quad . \tag{5.58}$$

The final recognition result in the multi-object scene $\boldsymbol{f}$ are the following object classes and poses

$$\begin{array}{ll} \text{first object} & (\kappa_1, \widehat{\boldsymbol{\phi}}_{\kappa_1}, \widehat{\boldsymbol{t}}_{\kappa_1}) \\ \text{second object} & (\kappa_2, \widehat{\boldsymbol{\phi}}_{\kappa_2}, \widehat{\boldsymbol{t}}_{\kappa_2}) \\ \vdots & \\ \text{last object} & (\kappa_{\widehat{i}}, \widehat{\boldsymbol{\phi}}_{\kappa_{\widehat{i}}}, \widehat{\boldsymbol{t}}_{\kappa_{\widehat{i}}}) \end{array} \quad . \tag{5.59}$$

In order to evaluate the recognition algorithm for multi-object scenes with context dependencies, not only the object classification result Ω_{κ_i} and the object localization result $(\widehat{\boldsymbol{\phi}}_{\kappa_i}, \widehat{\boldsymbol{t}}_{\kappa_i})$ have to be verified. The number $\widehat{i}$ of objects found in the scene $\boldsymbol{f}$ must also be checked (Section 6.5). Comparing experimental results in Section 6.5 and Section 6.6 one can see that including context modeling to the approach better classification rates are achieved.

5.6 Summary

In this chapter, the recognition phase of the system for object classification and localization was discussed. In the introduction to this chapter, a general description of all steps leading to the classification and localization result were presented and illustrated in Figure 5.1. First, an image from the real world environment is taken, preprocessed, and feature vectors in it are computed. Then the system starts one of the four recognition algorithms. The first algorithm deals with single-object scenes using one image, the second one works also with single-object scenes, but it uses multiple views of the object, the third one is applied for multi-object scenes without context dependencies, and the fourth one considers context dependencies in the multi-object scenes.

Section 5.1 described beginning steps of the recognition phase, i. e., image acquisition, image preprocessing, and feature extraction. The original test scene can be converted either to a square gray level image or to a square color image. In the gray level images, 2D local feature vectors are computed; while in the color images, 6D local feature vectors are determined (see Figure 5.2). Depending on the dimension of local feature vectors the statistical object models for gray level or color images are applied for further scene evaluation.

The algorithm for single-object scenes that uses only one image (one object view) for object classification and localization was explained in Section 5.2. The task of the algorithm is to find the class $\Omega_{\widehat{\kappa}}$, (or just its index $\widehat{\kappa}$) and the pose $(\widehat{\boldsymbol{\phi}}, \widehat{\boldsymbol{t}})$ of the object occurring in the test scene $\boldsymbol{f}$. In order to solve the problem, the object density values for all objects Ω_κ and many pose hypotheses $(\boldsymbol{\phi}_h, \boldsymbol{t}_h)$ have to be compared to each other. Section 5.2.1 described how to compute the object density value for the given test image $\boldsymbol{f}$, object class Ω_κ, and object pose hypothesis $(\boldsymbol{\phi}_h, \boldsymbol{t}_h)$. The recognition algorithm based on the maximum likelihood estimation was presented in Section 5.2.2. Finally, Section 5.2.3 extended the recognition algorithm in such a way that it is able to combine different resolution levels of the wavelet transformation for feature extraction.

Section 5.3 dealt with fusion of multiple views for object recognition. This algorithm expects still exactly one resulting object class $\Omega_{\widehat{\kappa}}$, but it uses a fusion of multiple different views $(\boldsymbol{f}_0, \boldsymbol{f}_1, \ldots, \boldsymbol{f}_n)$ of the object. The problem can be solved by maximization of the object state density (5.22). First, the object state density is recursively propagated (5.30) with the method described in Section 5.3.1. The practical realization of the fusion of multiple views with the condensation algorithm followed in Section 5.3.2.

Section 5.4 addressed the problem of object classification and localization in multi-object scenes without context modeling. In this case not only the resulting classes $(\Omega_{\kappa_1}, \Omega_{\kappa_2}, \ldots \Omega_{\kappa_{\widehat{\imath}}})$ of objects and their poses $((\widehat{\boldsymbol{\phi}}_{\kappa_1}, \widehat{\boldsymbol{t}}_{\kappa_1}), (\widehat{\boldsymbol{\phi}}_{\kappa_2}, \widehat{\boldsymbol{t}}_{\kappa_2}), \ldots, (\widehat{\boldsymbol{\phi}}_{\kappa_{\widehat{\imath}}}, \widehat{\boldsymbol{t}}_{\kappa_{\widehat{\imath}}}))$ have to be determined in the recognition phase. Since the number of objects $\widehat{\imath}$ in a scene is unknown, it must be also estimated. Section 5.4 started with a brief description of an approach for multi-object scenes, which is based on the so-called serial search algorithm. Finally, a simplification of this method was discussed. The new algorithm yields comparable results, but it is significantly faster.

Section 5.5 concerned also the problem of object classification and localization in multi-object scenes. However, in contrast to Section 5.4, it took into consideration context dependencies between objects, which were statistically modeled in the training phase (Section 4.4). Since there is no a-priori knowledge about the context Υ_ι (4.74) in which a test image $\boldsymbol{f}$ was taken, the algorithm is split into two phases. First, the context $\Upsilon_{\widehat{\imath}}$ is automatically determined for the test image $\boldsymbol{f}$. Then the approach classifies and localizes objects in the image $\boldsymbol{f}$ using the statistical context model $\mathcal{M}_{\widehat{\imath}}$ learned for this context $\Upsilon_{\widehat{\imath}}$ in the training phase (Section 4.4).

Chapter 6

Experiments and Results

Following the description of the training (Chapter 4) and the recognition phase (Chapter 5), the system is evaluated by experiments presented in this chapter. The following sections show that the using of color information for object representation (Section 4.3) as well as the context modeling (Section 4.4) bring a significant improvement of the recognition results in comparison to the approach introduced by Reinhold [Rei04].

The hardware and software used for experiments as well as the evaluation criteria are shortly described in Section 6.1. In [Rei04] the system was evaluated based on the DIROKOL image database [Rei01] (see Figure 3.4) with artificially created test images. In the present work the tests are performed using images with real heterogeneous background. For this purpose the image database for 3D object recognition in a real world environment (3D-REAL-ENV) was generated. Its description follows in Section 6.2.

As can be seen in Figure 5.1, there are four different recognition algorithms integrated into the system. The mathematical description of these algorithms can be found in Chapter 5. The present chapter aims to evaluate them. Section 6.3 shows results for the algorithm for single-object scenes introduced in Section 5.2. This recognition method uses only one image for object classification and localization. The fusion of multiple views realized with the condensation algorithm and described in detail in Section 5.3 finds its evaluation in Section 6.4. Section 6.5 presents the experimental results for the algorithm dealing with multi-object scenes without context dependencies explained in Section 5.4, while in Section 6.6 the recognition method for scenes with context dependencies (Section 5.5) is evaluated.

Section 6.7 closes this chapter summarizing and interpreting the classification and localization rates as well as other evaluation results for all recognition algorithms introduced in the present work.

6.1 Experimental Environment and Evaluation Criteria

In the present section, first, the hardware and software applied in the experimental phase are briefly described in Section 6.1.1. Subsequently, Section 6.1.2 defines the evaluation criteria used for experiments within the scope of this work.

6.1.1 Hardware and Software

For evaluation of the object recognition system images of real objects on real background were taken. According to Section 4.1.1 the training data can be collected either with a turntable and camera arm for laboratory applications (see Figure 4.2, left) or with a hand-held camera (see Figure 4.3, left). Test images with real background were acquired exclusively with the setup containing turntable and camera arm in a laboratory environment (see Figure 4.2, left). The following hardware was used for image acquisition:

- *Turntable and Camera Arm*

  Most of the images were taken using a special setup with turntable and camera arm depicted in Figure 4.2. The objects were put on the turntable ($0° \leq \phi_{\text{table}} < 360°$) and a robot arm with a camera (*Sony Firewire Camera DFW-VL-500*) was moved from vertical to horizontal ($0° \leq \phi_{\text{arm}} \leq 90°$). Both the turntable and the robot arm are moved from a *Nanotec Step Motor*, which is controlled by *Isel CNC-Controller C142-1*. The minimal increment of the step motor amounts to $1.8°$. However, there is a gear between the step motor and the turntable as well as between the step motor and the robot arm so that the minimal rotation angle for the turntable is equal to $\Delta\phi_{\text{table}} = 0.15°$ and the minimal rotation angle for the robot arm amounts to $\Delta\phi_{\text{arm}} = 0.03°$.

- *Hand-Held Camera*

  Some of the images were acquired by moving a hand-held camera around an object and showing it from different directions as illustrated in Figure 4.3. For this purpose the *Sony Digital Camera DSR-PD100A* was used.

Furthermore, the programs and scripts for experiments were executed on a common workstation equipped with Pentium 4, 2.66 GHz, and 512 MB RAM.

The system for appearance-based statistical object classification and localization was developed in object-oriented C++ programming language [Lip97] under Linux operating system, whereas some evaluation scripts were written using the practical extraction and report language (Perl). The basic software structure of the system was taken over from Reinhold [Rei04] and widely extended within the scope of the present work. Additionally, the system makes use of the following software components developed in object-oriented C++:

- *Class Hierarchy NIHCL [Gor90]*
- *Class Library HIPPOS for Image Processing [Pau91]*
- *Class Hierarchy STACCATO for Density Functions and Transformation Parameters [Hor96]*
- *Wavelet Transformation Implemented According to [Kra95]*
- *Downhill Simplex Algorithm Implemented According to [Pre90]*

Moreover, a graphical user interface (GUI) for demonstrating the object classification task in real world environment was developed using Perl/Tk programming language [Wal99].

6.1.2 Evaluation Criteria

The robustness of the system is expressed by the following score coefficients:

- *Classification Rate c_r*
  The object classes in the test images presented in Section 6.2.2 are a-priori known. However, the system determines them once again without this a-priori knowledge. Then, for each classification task, a comparison between the system classification result and the a-priori known class is made and the so-called *classification rate* is computed. The classification rate c_r is defined by

$$c_r = \frac{N_{\text{correct class}}}{N_{\text{correct class}} + N_{\text{incorrect class}}} \cdot 100\% \quad , \tag{6.1}$$

  where, as one can suppose, $N_{\text{correct class}}$ is the number of correct classification results, while $N_{\text{incorrect class}}$ is the number of incorrect classifications.

- *Localization Rate l_r*
  The object pose parameters defined according to Figure 1.2 are also known for all objects occurring in the test images. However, the system performs the object localization in the test scenes once again, whereas it is assumed that the object classes are known. Then, for each localization result, it has to be decided, if the localization was correct or not. A

localization result is counted as correct, if the error for internal translations $\boldsymbol{t}_{\text{int}} = (t_x, t_y)^{\text{T}}$ is not larger than 10 pixels

$$\varepsilon(t_x) \leq 10\,\text{pixels} \quad \textbf{and} \quad \varepsilon(t_y) \leq 10\,\text{pixels} \tag{6.2}$$

and the error for external translation $t_{\text{ext}} = t_z$ is not larger than 10%

$$\varepsilon(t_z) \leq 10\% \tag{6.3}$$

and the error for external rotations $\boldsymbol{\phi}_{\text{ext}} = (\phi_x, \phi_y)^{\text{T}}$ is not larger than 15°

$$\varepsilon(\phi_x) \leq 15^\circ \quad \textbf{and} \quad \varepsilon(\phi_y) \leq 15^\circ \tag{6.4}$$

and the error for internal rotation $\phi_{\text{int}} = \phi_z$ is not larger than 10°

$$\varepsilon(\phi_z) \leq 10^\circ \quad . \tag{6.5}$$

The number of correct localization results $N_{\text{correct pose}}$ and the number of incorrect localizations $N_{\text{incorrect pose}}$ determine the so-called *localization rate* l_r

$$l_r = \frac{N_{\text{correct pose}}}{N_{\text{correct pose}} + N_{\text{incorrect pose}}} \cdot 100\% \quad . \tag{6.6}$$

- *Object Number Determination Rate* d_r
  In the case of multi-object scenes, first, the system has to determine the number of objects occurring in the test scene. In order to express the robustness of this determination method the so-called *object number determination rate* d_r

  $$d_r = \frac{N_{\text{correct number}}}{N_{\text{correct number}} + N_{\text{incorrect number}}} \cdot 100\% \tag{6.7}$$

  is calculated. As one can suppose, $N_{\text{correct number}}$ is the number of correct determinations, while $N_{\text{incorrect number}}$ is the number of incorrect determinations.

The higher the score coefficients (c_r, l_r, and d_r), the higher is the performance of the object recognition system. Using the same image database and the same evaluation criteria, different systems for object recognition can be compared to each other.

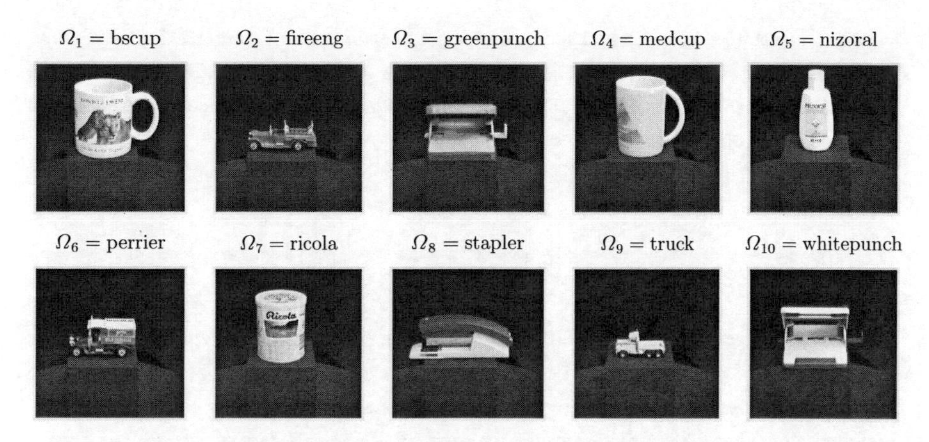

Figure 6.1: Ten objects of the 3D-REAL-ENV image database with their short names. First row from left: bank cup, toy fire engine, green puncher, siemens cup, nizoral bottle. Second row from left: toy passenger car, candy box, blue stapler, toy truck, white puncher.

6.2 Image Database

In the scope of the present work, the image database for 3D object recognition in a real world environment (3D-REAL-ENV) [Rei05, Grz05a, Grz05b] was acquired and the most experiments made for evaluation of the object recognition system were performed using this database. 3D-REAL-ENV image database consists of ten objects ($N_{\Omega} = 10$) depicted in Figure 6.1. The short names

$$\Omega = \left\{ \begin{array}{lllll} \text{bscup}, & \text{fireeng}, & \text{greenpunch}, & \text{medcup}, & \text{nizoral}, \\ \text{perrier}, & \text{ricola}, & \text{stapler}, & \text{truck}, & \text{whitepunch} \end{array} \right\} \quad (6.8)$$

for the object classes $\Omega_{\kappa=1,\ldots,10}$ are used in this form in the following. The pose (see Figure 1.2) of 3D-REAL-ENV objects is defined with internal translations $\boldsymbol{t}_{\text{int}} = (t_x, t_y)^{\text{T}}$ and external rotation parameters $\boldsymbol{\phi}_{\text{ext}} = (\phi_x, \phi_y)^{\text{T}}$. The internal rotation $\phi_{\text{int}} = \phi_z$ as well as the external translation $t_{\text{ext}} = t_z$ (scaling) are not taken into consideration in the original 3D-REAL-ENV image database. However, they can be generated in some postprocessing steps depending on the application. The image acquisition process was made under three different illumination conditions

$$I_{\text{lum}} \in \{\text{bright}, \text{middle}, \text{dark}\} \quad (6.9)$$

with the turntable and camera arm for laboratory applications (see Figure 4.2, left) and delivered RGB images sized 640×480 pixels, whereas for experiments, the images were resized to 256×256 pixels. The properties of the training images are discussed in Section 6.2.1. Section 6.2.2 introduces three types of test images, namely test images with homogeneous background, test images with less heterogeneous background, and test images with more heterogeneous background.

6.2.1 Training Images

The training images were taken on dark background from 1680 different viewpoints, whereas for all object classes $\Omega_{\kappa=1,\ldots,10}$ the same points of view were used. From all viewpoints, objects were acquired twice, i. e., under two different illumination conditions $I_{\text{lum}} \in \{\text{bright}, \text{dark}\}$. Hence, there are altogether 3360 training images for each 3D-REAL-ENV object class. The objects were put on a turntable moving around ($0° \leq \phi_{\text{table}} < 360°$) and a robot arm with a camera was moved from vertical to horizontal ($0° \leq \phi_{\text{arm}} \leq 90°$). The movement of the camera arm ϕ_{arm} corresponds to the first external rotation ϕ_x, while the turntable ϕ_{table} simulates the second external rotation parameter ϕ_y. The angle between two adjacent steps of the turntable and of the camera arm amounts to $4.5°$, which yields the following set of training views

$$\boldsymbol{\phi}_{\text{ext}} = (\phi_x, \phi_y)^{\mathrm{T}} \in \left\{ \begin{array}{llll} (0.0, 0.0)^{\mathrm{T}}, & (4.5, 0.0)^{\mathrm{T}}, & \ldots & (90.0, 0.0)^{\mathrm{T}}, \\ (0.0, 4.5)^{\mathrm{T}}, & (4.5, 4.5)^{\mathrm{T}}, & \ldots & (90.0, 4.5)^{\mathrm{T}}, \\ (0.0, 9.0)^{\mathrm{T}}, & (4.5, 9.0)^{\mathrm{T}}, & \ldots & (90.0, 9.0)^{\mathrm{T}}, \\ \vdots & \vdots & & \vdots \\ (0.0, 355.5)^{\mathrm{T}}, & (4.5, 355.5)^{\mathrm{T}}, & \ldots & (90.0, 355.5)^{\mathrm{T}} \end{array} \right\} \quad . \tag{6.10}$$

The external rotations ϕ_x and ϕ_y are given in degrees $[°]$. The number of training views given above can now be confirmed as follows

$$\frac{360}{4.5} \times \left(\frac{90}{4.5} + 1 \right) = 1680 \quad . \tag{6.11}$$

The total number of training images in 3D-REAL-ENV database is, therefore,

$$1680 \times 2 \times 10 = 33600 \quad . \tag{6.12}$$

Moving the turntable and the camera arm objects translate within the image plane, which results in varying internal translation parameters $\boldsymbol{t}_{\text{int}} = (t_x, t_y)^{\mathrm{T}}$. The internal translations for the

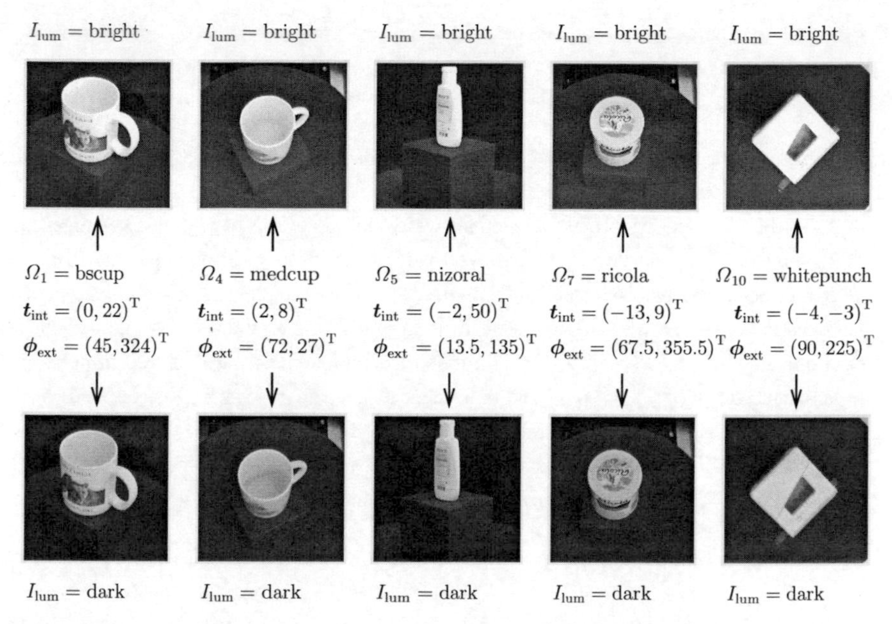

Figure 6.2: Ten example training images of five object classes {bscup, medcup, nizoral, ricola, whitepunch} taken under two boundary illumination conditions $I_{\text{lum}} \in \{\text{bright}, \text{dark}\}$ with specified object pose parameters, i. e., $\boldsymbol{t}_{\text{int}} = (t_x, t_y)^{\text{T}}$ given in pixels and $\boldsymbol{\phi}_{\text{ext}} = (\phi_x, \phi_y)^{\text{T}}$ given in degrees.

objects in training images are determined manually after acquisition. Figure 6.2 presents ten example training images of five object classes {bscup, medcup, nizoral, ricola, whitepunch} taken under two boundary illumination conditions $I_{\text{lum}} \in \{\text{bright}, \text{dark}\}$ with specified object pose parameters, i. e., $\boldsymbol{t}_{\text{int}} = (t_x, t_y)^{\text{T}}$ given in pixels and $\boldsymbol{\phi}_{\text{ext}} = (\phi_x, \phi_y)^{\text{T}}$ given in degrees.

6.2.2 Test Images

For the testing of the system for object classification and localization, the ten objects illustrated in Figure 6.1 were recorded from 288 different viewpoints under the middle illumination condition ($I_{\text{lum}} = \text{middle}$), whereas for all object classes $\Omega_{\kappa=1,\ldots,10}$ the same points of view were taken into account. There are three types of test images, namely test images with homogeneous background, test images with less heterogeneous background, and test images with more heterogeneous background. These types are denoted by

$$T_{\text{ype}} \in \{\text{hom}, \text{less}, \text{more}\} \quad . \tag{6.13}$$

Test scenes of the first type ($T_{\text{ype}} = \text{hom}$) were taken on dark homogeneous background, while for the acquisition of test images with heterogeneous background ($T_{\text{ype}} \in \{\text{less}, \text{more}\}$) more than 200 different heterogeneous backgrounds were applied. In scenes with less heterogeneous background ($T_{\text{ype}} = \text{less}$) the objects are easier to distinguish from the background than in images with more heterogeneous background ($T_{\text{ype}} = \text{more}$). Some example test images of all types can be found in Figure 6.3. Similarly to the acquisition of the training images, the objects were put on a turntable ($0° \leq \phi_{\text{table}} < 360°$) and a robot arm with a camera was moved from vertical to horizontal ($0° \leq \phi_{\text{arm}} \leq 90°$). The movement of the camera arm ϕ_{arm} corresponds to the first external rotation ϕ_x, while the rotation of the turntable ϕ_{table} simulates the second external rotation parameter ϕ_y. The angle between two adjacent steps of the turntable and of the camera arm amounts to $11.25°$ for test images, which yields the following set of test views

$$\boldsymbol{\phi}_{\text{ext}} = (\phi_x, \phi_y)^{\text{T}} \in \left\{ \begin{array}{llll} (0.0, 0.0)^{\text{T}}, & (11.25, 0.0)^{\text{T}}, & \ldots & (90.0, 0.0)^{\text{T}}, \\ (0.0, 11.25)^{\text{T}}, & (11.25, 11.25)^{\text{T}}, & \ldots & (90.0, 11.25)^{\text{T}}, \\ (0.0, 22.5)^{\text{T}}, & (11.25, 22.5)^{\text{T}}, & \ldots & (90.0, 22.5)^{\text{T}}, \\ \vdots & \vdots & & \vdots \\ (0.0, 348.75)^{\text{T}}, & (11.25, 348.75)^{\text{T}}, & \ldots & (90.0, 348.75)^{\text{T}} \end{array} \right\} . \tag{6.14}$$

Hence, the number of test viewpoints mentioned above can be confirmed by

$$\frac{360}{11.25} \times \left(\frac{90}{11.25} + 1 \right) = 288 \quad . \tag{6.15}$$

Therefore, the total number of test images in 3D-REAL-ENV database computes as follows

$$288 \times 3 \times 10 = 8640 \quad , \tag{6.16}$$

whereas 2880 of them are test images with homogeneous background ($T_{\text{ype}} = \text{hom}$) and 5760 of them are test scenes with real heterogeneous background ($T_{\text{ype}} \in \{\text{less}, \text{more}\}$). Comparing the list of training viewpoints (6.10) to the list of test views (6.14) one can see that, in general, they are different. Moreover, the illumination in the test scenes is different from the illumination in the training images (see Figure 6.4). Due to all these properties, the object classification and localization task using the 3D-REAL-ENV image database becomes very difficult.

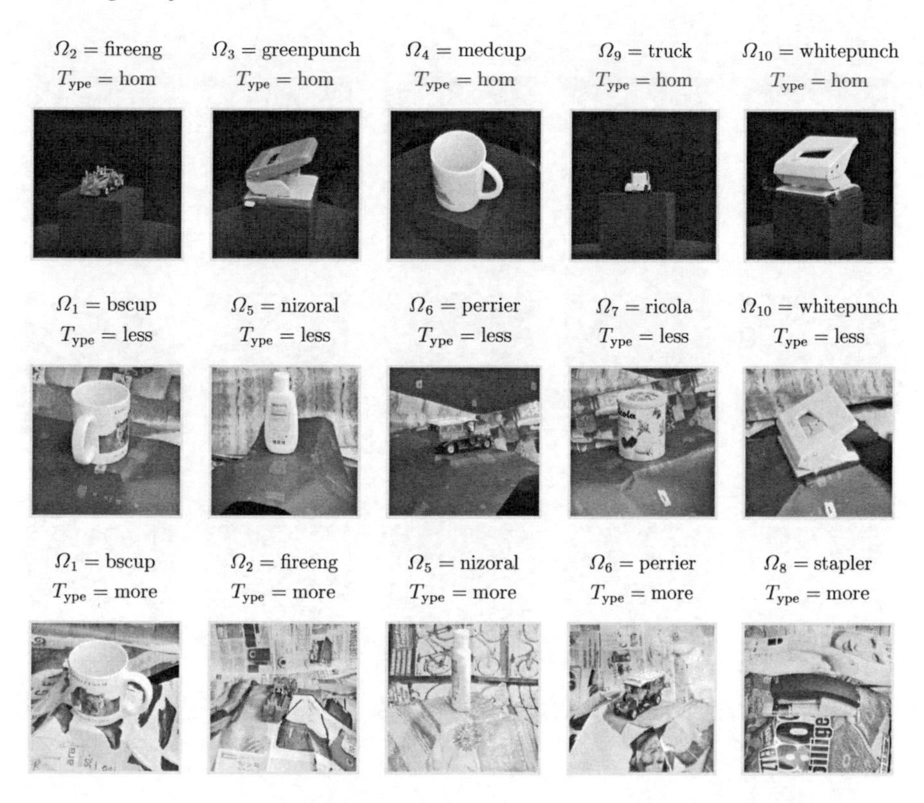

Figure 6.3: Example test images of all types $T_{\text{ype}} \in \{\text{hom}, \text{less}, \text{more}\}$. In the first row images with homogeneous background ($T_{\text{ype}} = \text{hom}$), in the second row images with less heterogeneous background ($T_{\text{ype}} = \text{less}$), and in the third row images with more heterogeneous background ($T_{\text{ype}} = \text{more}$) can be seen.

6.3 Single-Object Scenes

In this section, experiments made for images of single-objects distinguishable with only one view are described and results obtained with this approach, introduced in Section 5.2, are presented. In other words, exactly one resulting object class $\Omega_{\widehat{\kappa}}$ is expected and exactly one test scene $\boldsymbol{f}$ is regarded for determination of this object $\Omega_{\widehat{\kappa}}$ and its pose $(\widehat{\boldsymbol{\phi}}_{\widehat{\kappa}}, \widehat{\boldsymbol{t}}_{\widehat{\kappa}})$. Experiments discussed in this section were performed exclusively with the 3D-REAL-ENV image database introduced

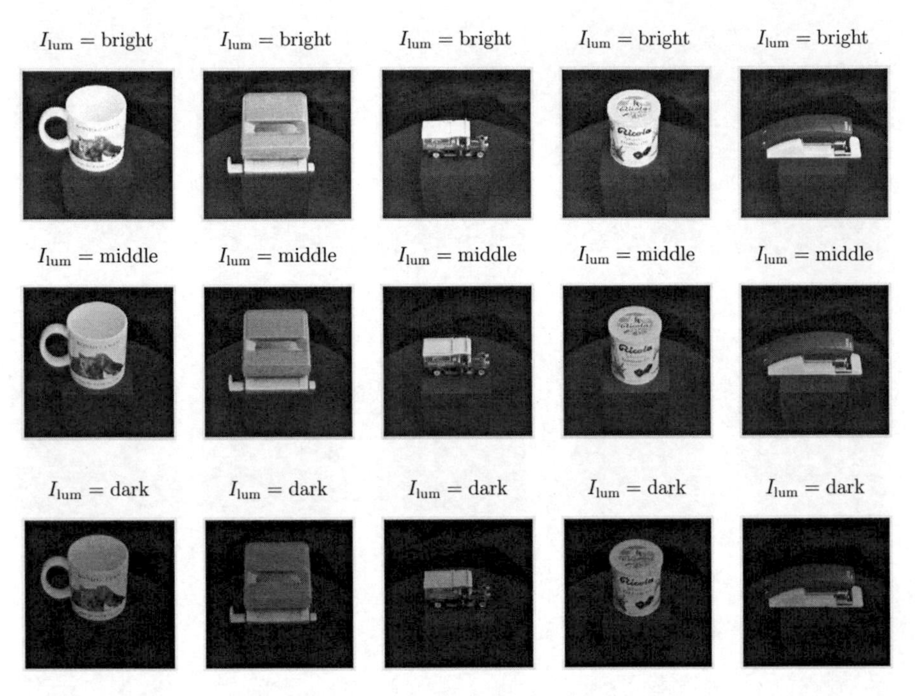

Figure 6.4: Five example objects taken under three different illumination conditions. First row: example training images taken under ($I_{\mathrm{lum}} = \mathrm{bright}$). Second row: corresponding test images taken under ($I_{\mathrm{lum}} = \mathrm{middle}$). Third row: corresponding training images taken under ($I_{\mathrm{lum}} = \mathrm{dark}$).

in Section 6.2. The practical realization of the training phase for the 3D-REAL-ENV objects depicted in Figure 6.1 is discussed in Section 6.3.1. Section 6.3.2 presents the classification and localization rates obtained for the 3D-REAL-ENV test images with the recognition algorithm introduced in Section 5.2.

6.3.1 Training

In the preprocessing stage for the training phase, all 33600 training images from 3D-REAL-ENV (Section 6.2.1) were transformed into square gray level images and square RGB images sized 256×256 pixels. Subsequently, the system performed the statistical object modeling

Number of Training Views and Images for 3D-REAL-ENV Image Database						
Distance of Training Views [°]	4.5	9.0	13.5	18.0	22.5	27.0
Number of Training Views	1680	440	189	120	80	56
Number of Training Images	33600	8800	3780	2400	1600	1120

Table 6.1: The number of 3D-REAL-ENV training views decreases with their increasing distance. Automatically, the total number of images used for training becomes lower, too.

(Chapter 4) for all object classes $\Omega_{\kappa=1,\ldots,10}$ contained in 3D-REAL-ENV image database (see Figure 6.1). First, gray level modeling for different densities of training views (altogether 6 density levels) was applied with the method discussed in Section 4.2. Second, RGB images for the training of object models (Section 4.3) were used, whereas the number of training views was also varying. Finally, for each object class Ω_κ from 3D-REAL-ENV image database, the following 12 statistical object models were created

$$\Omega_\kappa \Rightarrow \left\{ \begin{array}{cccccc} \mathcal{M}_{\kappa,04.5,g}, & \mathcal{M}_{\kappa,09.0,g}, & \mathcal{M}_{\kappa,13.5,g}, & \mathcal{M}_{\kappa,18.0,g}, & \mathcal{M}_{\kappa,22.5,g}, & \mathcal{M}_{\kappa,27.0,g}, \\ \mathcal{M}_{\kappa,04.5,c}, & \mathcal{M}_{\kappa,09.0,c}, & \mathcal{M}_{\kappa,13.5,c}, & \mathcal{M}_{\kappa,18.0,c}, & \mathcal{M}_{\kappa,22.5,c}, & \mathcal{M}_{\kappa,27.0,c} \end{array} \right\} . \quad (6.17)$$

For example, $\mathcal{M}_{\kappa,09.0,g}$ denotes a statistical object model learned for the object class Ω_κ using gray level images g (Section 4.2), where the distance of training views amounts to 9°, while, e. g., $\mathcal{M}_{\kappa,13.5,c}$ is a statistical object model created for Ω_κ using color images c (Section 4.3), where the distance of training views is 13.5°. The number of 3D-REAL-ENV views used for training decreases with their increasing distance, which can be seen in Table 6.1. If the angle between two adjacent training viewpoints amounts to, e. g., 18°, the following set of views is taken into consideration for training

$$\phi_{\text{ext}} = (\phi_x, \phi_y)^{\text{T}} \in \left\{ \begin{array}{llll} (0.0, 0.0)^{\text{T}}, & (18.0, 0.0)^{\text{T}}, & \ldots & (90.0, 0.0)^{\text{T}}, \\ (0.0, 18.0)^{\text{T}}, & (18.0, 18.0)^{\text{T}}, & \ldots & (90.0, 18.0)^{\text{T}}, \\ (0.0, 36.0)^{\text{T}}, & (18.0, 36.0)^{\text{T}}, & \ldots & (90.0, 36.0)^{\text{T}}, \\ \vdots & \vdots & & \vdots \\ (0.0, 342.0)^{\text{T}}, & (18.0, 342.0)^{\text{T}}, & \ldots & (90.0, 342.0)^{\text{T}} \end{array} \right\} . \tag{6.18}$$

The number of training views given in Table 6.1 for this case can be confirmed as follows

$$\frac{360}{18} \times \left(\frac{90}{18} + 1 \right) = 120 \tag{6.19}$$

and, accordingly, the total number of training images in this case is equal to

$$120 \times 2 \times 10 = 2400 \quad . \tag{6.20}$$

On the one hand, the objective of object modeling in the training phase is a satisfying description of objects in terms of recognition task. This is evaluated with the classification and localization rates in Section 6.3.2. On the other hand, a significant compression of the data amount in comparison to the size of training images is expected. Table 6.2 compares the total training data size of 3D-REAL-ENV before and after object modeling.

6.3.2 Recognition

The same as for the training images (Section 6.3.1), all 8640 test images from 3D-REAL-ENV introduced in Section 6.2.2 were preprocessed into square gray level images and square RGB images sized 256×256 pixels. Subsequently, the system performed the algorithm for object classification and localization described in Section 5.2 for all of them, i. e., for 2880 test images with homogeneous background, 2880 test images with less heterogeneous background, and 2880 test images with more heterogeneous background. As one can suppose, for gray level test images, statistical object models created using gray level training images (Section 4.2) were applied

$$\left\{ \mathcal{M}_{\kappa,04.5,g}, \ \mathcal{M}_{\kappa,09.0,g}, \ \mathcal{M}_{\kappa,13.5,g}, \ \mathcal{M}_{\kappa,18.0,g}, \ \mathcal{M}_{\kappa,22.5,g}, \ \mathcal{M}_{\kappa,27.0,g} \right\} \quad , \tag{6.21}$$

Training Data Size for 3D-REAL-ENV Image Database			
Distance of Views [°]	Training Images [MB]	Gray Level Models [MB]	Color Models [MB]
4.5	30966	5.6	12.9
9.0	8110	5.5	12.8
13.5	3484	5.5	12.7
18.0	2212	5.3	12.4
22.5	1475	5.3	12.3
27.0	1032	5.3	12.4

Table 6.2: Comparison of training data size for all 3D-REAL-ENV object classes before and after object modeling. Statistical modeling in the training phase significantly decreases the size of data describing objects.

while for color test images, results of color modeling presented in Section 4.3 were taken into consideration

$$\left\{ \mathcal{M}_{\kappa,04.5,c}, \quad \mathcal{M}_{\kappa,09.0,c}, \quad \mathcal{M}_{\kappa,13.5,c}, \quad \mathcal{M}_{\kappa,18.0,c}, \quad \mathcal{M}_{\kappa,22.5,c}, \quad \mathcal{M}_{\kappa,27.0,c} \right\} \quad . \tag{6.22}$$

The classification and localization rates for this series of experiments calculated according to evaluation criteria defined in Section 6.1.2 are presented in Table 6.3. As can be seen, the distance of training views varies from $4.5°$ to $27°$ in 5 steps. Figure 6.5 instead compares the recognition rates obtained for gray level and color modeling for all three types of test images ($T_{\mathrm{ype}} \in \{\mathrm{hom}, \mathrm{less}, \mathrm{more}\}$) presenting them as functions of the training view distance. The color modeling brings a significant improvement of the classification and localization rates for test images with more heterogeneous background. For scenes with homogeneous and less heterogeneous background, the recognition algorithm works very well using gray level modeling, and it is not necessary to consider the color information of objects. The object recognition takes 3.6s in one gray level image and 7s in one color image on a workstation equipped with Pentium 4, 2.66 GHz, and 512 MB RAM.

Recognition Rates for 3D-REAL-ENV Image Database								
Distance of Training Views [°]	Type of Object Modeling	Classification Rate [%]			Localization Rate [%]			
		Hom. Back.	Less Het. Back.	More Het. Back.	Hom. Back.	Less Het. Back.	More Het. Back.	
4.5	Gray	100	92.2	54.1	99.1	80.9	69.0	
	Color	100	88.0	82.3	98.5	77.8	73.6	
9.0	Gray	100	92.4	55.4	98.7	80.0	67.2	
	Color	100	88.3	81.2	98.2	76.4	72.1	
13.5	Gray	99.4	89.7	56.2	96.9	78.6	65.4	
	Color	99.6	82.7	80.3	94.9	68.4	66.6	
18.0	Gray	99.9	89.2	55.1	96.6	71.4	54.5	
	Color	97.3	80.6	68.6	94.3	64.9	60.7	
22.5	Gray	99.4	86.0	52.8	94.5	60.7	38.6	
	Color	94.7	74.8	59.2	89.4	52.2	46.2	
27.0	Gray	96.5	69.4	54.4	83.8	49.9	32.8	
	Color	93.8	53.6	50.2	78.3	35.8	35.6	

Table 6.3: Classification and localization rates obtained for 3D-REAL-ENV image database with gray level and color modeling. The distance of training views varies from 4.5° to 27° in 5 steps. For experiments, 2880 test images with homogeneous, 2880 test images with less heterogeneous, and 2880 images with more heterogeneous background were used.

6.4 Fusion of Multiple Views

The fusion of multiple views realized with the condensation algorithm [Grz03] and described in detail in Section 5.3 finds its evaluation in this section. The experiments were performed for the so-called *fusion image database* with images of single-objects, which are not distinguishable from some points of view. In other words, still exactly one resulting object class $\Omega_{\widehat{\kappa}}$ is expected, but a fusion of multiple different object views

$$\langle \boldsymbol{f} \rangle_n = (\boldsymbol{f}_0, \boldsymbol{f}_1, \ldots, \boldsymbol{f}_n) \qquad (6.23)$$

is considered for classification. The fusion image database consists of the eight objects depicted in Figure 6.6. The practical realization of the statistical modeling for these objects is discussed in Section 6.4.1. Section 6.4.2 compares the classification rates obtained using the fusion algorithm (Section 5.3) to the classification results achieved with the simple algorithm for single-object

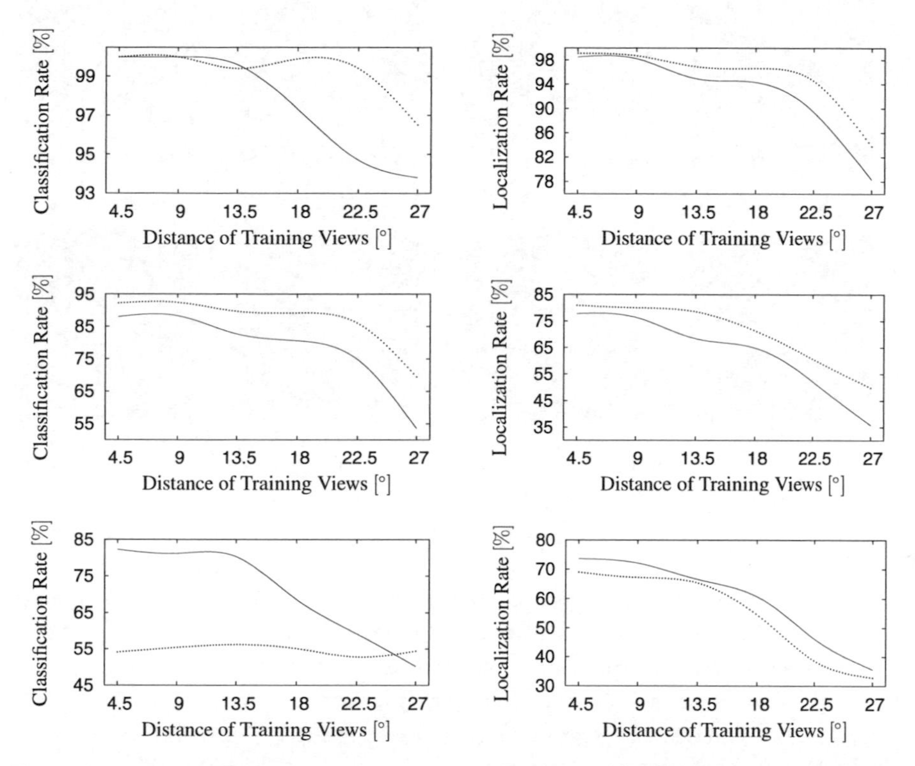

Figure 6.5: Classification and localization rates obtained for 3D-REAL-ENV image database depending on the distance of training views for 2880 test images with homogeneous (first row), 2880 test images with less heterogeneous (second row), and 2880 test images with more heterogeneous background (third row). (— color modeling; ··· gray level modeling).

scenes (Section 5.2). Furthermore, it presents the localization rates obtained for this database.

6.4.1 Training

Training images of the objects depicted in Figure 6.6 were taken with the same method and configuration as training images of the 3D-REAL-ENV objects (Section 6.2.1). They were acquired

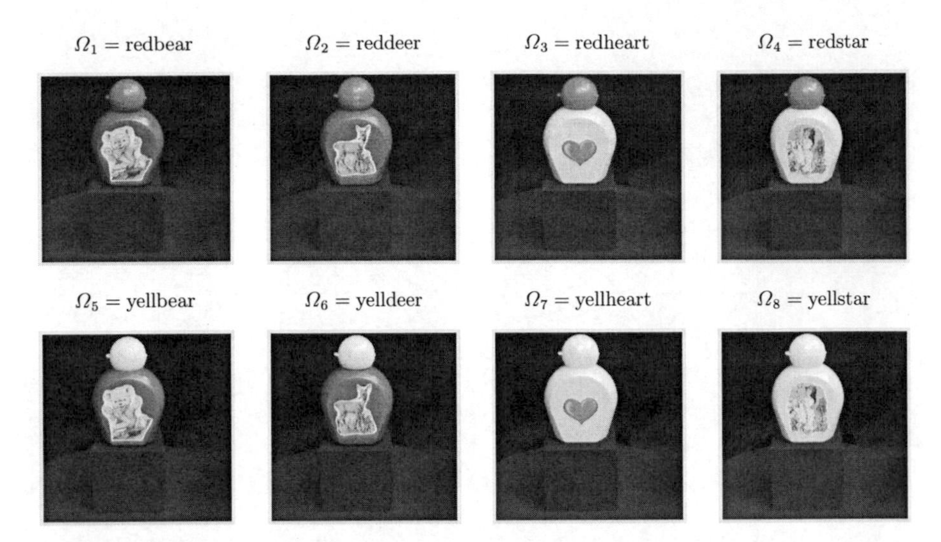

Figure 6.6: Eight objects used for fusion of multiple views with their short names. The bottles are pairwise not to distinguish from behind (see Figure 1.4).

on dark background from 1680 different viewpoints under two different illumination conditions $I_{\text{lum}} \in \{\text{bright}, \text{dark}\}$. The objects were put on a turntable moving around ($0° \leq \phi_{\text{table}} < 360°$) and a robot arm with a camera was moved from vertical to horizontal ($0° \leq \phi_{\text{arm}} \leq 90°$). The movement of the camera arm ϕ_{arm} corresponds to the first external rotation ϕ_x, while the turntable ϕ_{table} simulates the second external rotation parameter ϕ_y. The angle between two adjacent steps of the turntable and of the camera arm amounts to $4.5°$, which yields the following set of training views

$$\boldsymbol{\phi}_{\text{ext}} = (\phi_x, \phi_y)^{\mathrm{T}} \in \left\{ \begin{array}{llll} (0.0, 0.0)^{\mathrm{T}}, & (4.5, 0.0)^{\mathrm{T}}, & \ldots & (90.0, 0.0)^{\mathrm{T}}, \\ (0.0, 4.5)^{\mathrm{T}}, & (4.5, 4.5)^{\mathrm{T}}, & \ldots & (90.0, 4.5)^{\mathrm{T}}, \\ \vdots & \vdots & & \vdots \\ (0.0, 355.5)^{\mathrm{T}}, & (4.5, 355.5)^{\mathrm{T}}, & \ldots & (90.0, 355.5)^{\mathrm{T}} \end{array} \right\} . \quad (6.24)$$

Considering the two illuminations $I_{\text{lum}} \in \{\text{bright}, \text{dark}\}$ for all training views, there are altogether 3360 training images for each of the eight objects.

In the preprocessing stage for the training phase, all these 26880 training images were transformed into square gray level images sized 256×256 pixels. Subsequently, the system performed the statistical object modeling for gray level images as described in Section 4.2. Finally, for each object class Ω_κ visible in Figure 6.6 exactly one statistical object model $\mathcal{M}_{\kappa,04.5,g}$ was created. On the one hand, the objective of object modeling in the training phase is a satisfying description of objects in terms of recognition task. This is evaluated with the recognition rates in Section 6.4.2. On the other hand, a significant compression of the data amount in comparison to the size of training images is expected. In the case of objects presented in Figure 6.6 the total size of training images amounts to 24773 MB, while the total size of object models is 3.4 MB.

6.4.2 Recognition

For testing of the recognition algorithm described in Section 5.3 the eight objects illustrated in Figure 6.6 were recorded on homogeneous background (T_{ype} = hom) from 288 different viewpoints under the middle illumination condition (I_{lum} = middle), whereas for all object classes $\Omega_{\kappa=1,\ldots,8}$ the same points of view were taken into account. Hence, there are altogether 2304 test images in this case. Similarly to the acquisition of the training images, the objects were put on a turntable ($0° \leq \phi_{\text{table}} < 360°$) and a robot arm with a camera was moved from vertical to horizontal ($0° \leq \phi_{\text{arm}} \leq 90°$). The movement of the camera arm ϕ_{arm} corresponds to the first external rotation ϕ_x, while the rotation of the turntable ϕ_{table} simulates the second external rotation parameter ϕ_y. The angle between two adjacent steps of the turntable and of the camera arm amounts to $11.25°$ for test images, which yields the following set of test views

$$\phi_{\text{ext}} = (\phi_x, \phi_y)^{\mathrm{T}} \in \left\{ \begin{array}{llll} (0.0, 0.0)^{\mathrm{T}}, & (11.25, 0.0)^{\mathrm{T}}, & \ldots & (90.0, 0.0)^{\mathrm{T}}, \\ (0.0, 11.25)^{\mathrm{T}}, & (11.25, 11.25)^{\mathrm{T}}, & \ldots & (90.0, 11.25)^{\mathrm{T}}, \\ (0.0, 22.5)^{\mathrm{T}}, & (11.25, 22.5)^{\mathrm{T}}, & \ldots & (90.0, 22.5)^{\mathrm{T}}, \\ \vdots & \vdots & & \vdots \\ (0.0, 348.75)^{\mathrm{T}}, & (11.25, 348.75)^{\mathrm{T}}, & \ldots & (90.0, 348.75)^{\mathrm{T}} \end{array} \right\} . \quad (6.25)$$

Comparing the list of training viewpoints (6.24) to the list of test views (6.25) one can see that, in general, they are different. Moreover, illumination in the test scenes is different from illumination in the training images.

First, experiments for the test images described above were performed without using the fusion algorithm, i. e., the simple recognition algorithm for single-object scenes presented in Section 5.2 was applied. The classification rate obtained for all 2304 test scenes amounts to 68.1% in this case. Since the objects depicted in Figure 6.6 are pairwise not distinguishable from

Confusion Matrix for Classification without Fusion Algorithm								
	Ω_1	Ω_2	Ω_3	Ω_4	Ω_5	Ω_6	Ω_7	Ω_8
$\Omega_1 = \mathrm{redbear}$	28.1	71.9	0	0	0	0	0	0
$\Omega_2 = \mathrm{reddeer}$	15.6	84.4	0	0	0	0	0	0
$\Omega_3 = \mathrm{redheart}$	0	0	45.8	54.2	0	0	0	0
$\Omega_4 = \mathrm{redstar}$	0	0	3.1	96.9	0	0	0	0
$\Omega_5 = \mathrm{yellbear}$	0	0	0	0	56.6	43.4	0	0
$\Omega_6 = \mathrm{yelldeer}$	0	0	0	0	10.4	89.6	0	0
$\Omega_7 = \mathrm{yellheart}$	0	0	0	0	0	0	93.1	6.9
$\Omega_8 = \mathrm{yellstar}$	0	0	0	0	0	0	49.3	50.7

Table 6.4: Confusion matrix for classification of objects depicted in Figure 6.6 without using the fusion algorithm. All numbers are given in [%]. For example, in the row beginning with "$\Omega_4 = \mathrm{redstar}$" classification results for test images with this object ("redstar") are presented. The total classification rate for all 2304 test scenes amounts to 68.1%.

behind (see Figure 1.4), the system confuses them often with each other, which is quantitatively expressed in Table 6.4.

In the second series of experiments for the test images described at the beginning of this section, the algorithm for fusion of multiple views introduced in Section 5.3 was applied. The total classification rate obtained for all 2304 test scenes amounts to 80.2% in this case. As can be seen, it is significantly higher than the classification rate (68.1%) for experiments without using the fusion algorithm. Even for the objects depicted in Figure 6.6, which are pairwise not distinguishable from behind (see Figure 6.6), the object confusion occurs relatively seldom. The so-called confusion matrix for classification is presented in Table 6.5 for this case.

Since the localization task is defined as determination of the object pose, whereas the object class is a-priori known (Section 1.1), the fusion algorithm presented in Section 5.3 is not applied in this case. The localization of the objects depicted in Figure 6.6 is performed with the simple algorithm for single-object scenes (Section 5.2). The total localization rate obtained for all 2304 test images from the fusion database considering the evaluation criteria defined in Section 6.1.2 amounts to 95.1%. Additionally, in Table 6.6, the localization rates for all eight objects from the fusion image database (see Figure 6.6) are compared.

Confusion Matrix for Classification with Fusion Algorithm								
	Ω_1	Ω_2	Ω_3	Ω_4	Ω_5	Ω_6	Ω_7	Ω_8
$\Omega_1 = \text{redbear}$	2.4	97.6	0	0	0	0	0	0
$\Omega_2 = \text{reddeer}$	2.1	97.9	0	0	0	0	0	0
$\Omega_3 = \text{redheart}$	0	0	46.5	53.5	0	0	0	0
$\Omega_4 = \text{redstar}$	0	0	0	100	0	0	0	0
$\Omega_5 = \text{yellbear}$	0	0	0	0	98.6	1.4	0	0
$\Omega_6 = \text{yelldeer}$	0	0	0	0	1.7	98.3	0	0
$\Omega_7 = \text{yellheart}$	0	0	0	0	0	0	97.9	2.1
$\Omega_8 = \text{yellstar}$	0	0	0	0	0	0	0	100

Table 6.5: Confusion matrix for classification of objects depicted in Figure 6.6 with the fusion algorithm. All numbers are given in [%]. For example, in the row beginning with "$\Omega_4 = \text{redstar}$" classification results for test images with this object ("redstar") are presented. The total classification rate for all 2304 test scenes amounts to 80.2%.

Localization Rates for Fusion Image Database							
Ω_1	Ω_2	Ω_3	Ω_4	Ω_5	Ω_6	Ω_7	Ω_8
96.9%	97.6%	91.7%	94.4%	94.8%	93.8%	95.5%	95.8%

Table 6.6: Localization rates for objects from the fusion image database (see Figure 6.6). The total localization rate for all 2304 test scenes amounts to 95.1%.

6.5 Multi-Object Scenes

The experimental results for the algorithm dealing with multi-object scenes without context dependencies, which is theoretical described in Section 5.4, are presented in this section. The experiments were made on the 3D-REAL-ENV objects depicted in Figure 6.1. The approach assumes the uniform distribution of the a-priori occurrence probabilities for all object classes

$$p(\Omega_1) = p(\Omega_2) = \cdots = p(\Omega_\kappa) = \cdots = p(\Omega_{10}) \tag{6.26}$$

and does not consider the context models learned in Section 4.4. Furthermore, in contrast to Section 6.4, only one test scene is regarded for object classification and localization. In the recognition task for multi-object scenes, the classes and poses as well as the number of objects in a scene have to be determined. The practical realization of the training phase for this algorithm is discussed in Section 6.5.1, while Section 6.5.2 evaluates the object number determination as well as the object classification and localization for this approach.

6.5.1 Training

Although the algorithm in the present section concerns multi-object scenes, the training of the statistical object models was made separately for all 3D-REAL-ENV object classes with a similar configuration to the training phase for single-object scenes described in detail in Section 6.3.1. In the preprocessing stage for the training phase, all 33600 training images from 3D-REAL-ENV (Section 6.2.1) were transformed into square gray level images sized 256×256 pixels. Subsequently, the system performed the statistical object modeling for gray level images (Section 4.2) for all object classes $\Omega_{\kappa=1,\dots,10}$ contained in 3D-REAL-ENV image database (see Figure 6.1). Finally, for each 3D-REAL-ENV object class Ω_κ exactly one statistical object model $\mathcal{M}_{\kappa,04.5,g}$ was generated. On the one hand, the objective of object modeling in the training phase is a satisfying description of objects in terms of recognition task. This is evaluated with the object number determination rates as well as the classification and localization rates in Section 6.5.2. On the other hand, a significant compression of the data amount in comparison to the size of training images is expected. In the case of the 3D-REAL-ENV objects presented in Figure 6.1, the total size of training images amounts to 30966 MB, while the total size of object models is 5.6 MB.

6.5.2 Recognition

In the testing phase of the recognition algorithm introduced in Section 5.4 and evaluated in the present section, altogether 3240 gray level multi-object scenes sized 512×512 pixels were used. They were generated based on the 3D-REAL-ENV single-object test images described in detail in Section 6.2.2, and can be divided into three types, i. e., there are 1080 multi-object scenes with homogeneous ($T_{\text{ype}} = \text{hom}$), 1080 multi-object scenes with less heterogeneous ($T_{\text{ype}} = \text{less}$), and 1080 multi-object scenes with more heterogeneous ($T_{\text{ype}} = \text{more}$) background. Additionally, the 3D-REAL-ENV objects (see Figure 6.1) were assigned into three different contexts, namely the kitchen Υ_1

$$\{\text{bscup}, \text{medcup}, \text{ricola}\} \in \Upsilon_1 \quad , \tag{6.27}$$

the nursery Υ_2

$$\{\text{fireeng}, \text{perrier}, \text{truck}\} \in \Upsilon_2 \quad , \tag{6.28}$$

Object Number Determination Rates for Multi-Object Scenes		
Homogeneous Background	Less Heterogeneous Background	More Heterogeneous Background
100%	83.9%	43.2%

Table 6.7: Object number determination rates for multi-object scenes estimated for 1080 test images with homogeneous, 1080 test images with less heterogeneous, and 1080 test images with more heterogeneous background.

and the office Υ_3

$$\{\mathrm{greenpunch}, \mathrm{stapler}, \mathrm{whitepunch}\} \in \Upsilon_3 \quad . \tag{6.29}$$

Since the context dependencies between the objects are not taken into consideration in the present section (6.26), the object assignment to the contexts does not have any further meaning. It becomes meaningful in Section 6.6. For each image type ($T_{\mathrm{ype}} \in \{\mathrm{hom}, \mathrm{less}, \mathrm{more}\}$) and each context ($\Upsilon = \{\mathrm{kitchen}, \mathrm{nursery}, \mathrm{office}\}$) 120 one-object, 120 two-object, and 120 three-object scenes were created, whereas the viewpoints were chosen randomly from all 288 3D-REAL-ENV test views (6.14) and are different for all combinations of the test scenes. For example, two-object test scenes with homogeneous background ($T_{\mathrm{ype}} = \mathrm{hom}$) from the kitchen context ($\Upsilon_1 = \mathrm{kitchen}$) represent, in general, different viewpoints as the three-object test scenes with less heterogeneous background ($T_{\mathrm{ype}} = \mathrm{less}$) from the office context ($\Upsilon_3 = \mathrm{office}$). Examples of multi-object test scenes of all types ($T_{\mathrm{ype}} \in \{\mathrm{hom}, \mathrm{less}, \mathrm{more}\}$) and all contexts ($\Upsilon = \{\mathrm{kitchen}, \mathrm{nursery}, \mathrm{office}\}$) are depicted in Figure 6.7. As one can see, even by humans it is extremely difficult to recognize the 3D-REAL-ENV objects, especially in the test images with more heterogeneous background.

Since the number of objects in a multi-object scene is not assumed to be a-priori known, the system determines it first. In order to express the robustness of this determination method, the so-called object number determination rate d_r, which was introduced in Section 6.1.2 (6.7), is calculated in the present section. In Table 6.7, the robustness of the object number determination is quantitatively expressed for all types of test images ($T_{\mathrm{ype}} \in \{\mathrm{hom}, \mathrm{less}, \mathrm{more}\}$).

For object classification, the system assumes that the number of objects in a multi-object scene is known. Moreover, the object localization is made for a-priori known object classes. Considering the configuration of the multi-object test scenes described in Section 6.5.2, there are 1080 three-object, 1080 two-object, and 1080 one-object images. Therefore, altogether 6480

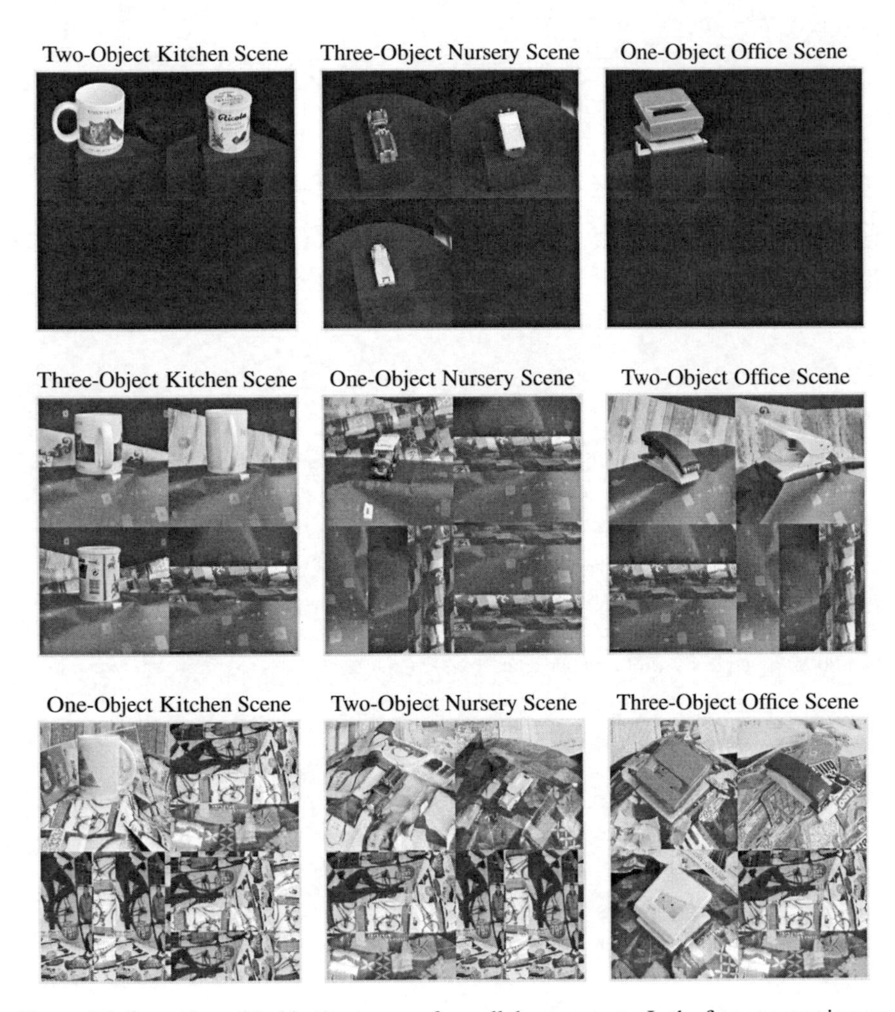

Figure 6.7: Example multi-object test scenes from all three contexts. In the first row, test images with homogeneous; in the second row, test images with less heterogeneous; and in the third row, test images with more heterogeneous background can be seen. Even by humans, it is extremely difficult to find the 3D-REAL-ENV objects, especially in the images with more heterogeneous background.

Classification and Localization Rates for Multi-Object Scenes			
	Homogeneous Background	Less Heterogeneous Background	More Heterogeneous Background
Classification	100%	91.9%	62.9%
Localization	99.7%	81.7%	58.1%

Table 6.8: Classification and localization rates for multi-object scenes estimated for 2160 recognition tasks with homogeneous, 2160 recognition tasks with less heterogeneous, and 2160 recognition tasks with more heterogeneous background.

objects have to be classified and localized in these images, whereas 2160 of them are placed on homogeneous, 2160 on less heterogeneous, and 2160 on more heterogeneous background. Table 6.8 presents the classification and localization rates for all multi-object test images (see Figure 6.7) achieved with the recognition approach introduced in Section 5.4.

6.6 Scenes with Context Dependencies

Experiments and results for the recognition algorithm dealing with multi-object scenes and considering context dependencies between objects are discussed in this section. They were performed using the 3D-REAL-ENV objects depicted in Figure 6.1. In contrast to Section 6.5, in this section the a-priori occurrence probabilities $p(\Omega_\kappa)$ for the 3D-REAL-ENV object classes are not assumed to be equal

$$p(\Omega_1) \neq p(\Omega_2) \neq \cdots \neq p(\Omega_\kappa) \neq \cdots \neq p(\Omega_{10}) \quad . \tag{6.30}$$

Instead of that, they are learned in the training phase as shown in Section 4.4 and stored in the context models $\mathcal{M}_\iota$. In the recognition phase, which is performed with the approach described in Section 5.5, the classes and poses as well as the number of objects in a scene have to be determined. The practical realization of the training phase for this algorithm is discussed in Section 6.6.1, while Section 6.6.2 evaluates the object number determination as well as the object classification and localization for this approach.

6.6.1 Training

Training of the 3D-REAL-ENV objects (see Figure 6.1) was performed with the same method and configuration as in Section 6.5.1. First, all 33600 training images from 3D-REAL-ENV (Section 6.2.1) were transformed into square gray level images sized 256×256 pixels. Second, the learning algorithm for gray level images introduced in Section 4.2 was applied for all 3D-REAL-ENV object classes. Finally, for each 3D-REAL-ENV object class Ω_κ exactly one statistical object model $\mathcal{M}_{\kappa,04.5,g}$ was generated. Due to the statistical object modeling, a significant data compression was achieved. The total size of training images amounts to 30966 MB, while the total size of object models is 5.6 MB for the 3D-REAL-ENV image database in this configuration.

In addition to the statistical object modeling for the 3D-REAL-ENV image database, context models $\mathcal{M}_\iota$ were also trained (see Section 4.4) as a part of experiments described in this section. First, the set of following contexts

$$\Upsilon = \{\text{kitchen}, \text{nursery}, \text{office}\} \tag{6.31}$$

was introduced. Then, for each context $N_\iota = 100$ images were taken with a hand-held camera (see Figure 4.3, left) from random viewpoints. Subsequently, it was manually counted, which of the 3D-REAL-ENV objects and how often occurred in these images, whereas with $N_{\iota,\kappa}$ the number is denoted, how often the object Ω_κ occurred in the context Υ_ι. Finally, the a-priori occurrence probability for the object Ω_κ in the context Υ_ι was learned as

$$p_\iota(\Omega_\kappa) = \eta_\iota N_{\iota,\kappa} \quad , \tag{6.32}$$

where η_ι is a normalization factor so that

$$\eta_\iota \left\{ p_\iota(\Omega_1) + p_\iota(\Omega_2) + \ldots + p_\iota(\Omega_{10}) \right\} = 1 \quad . \tag{6.33}$$

The learned values of the a-priori occurrence probabilities $p_\iota(\Omega_\kappa)$ for all 3D-REAL-ENV objects $\Omega_{\kappa=1,\ldots,10}$ in all contexts $\Upsilon_{\iota=1,2,3}$ are presented in Table 6.9.

6.6.2 Recognition

The algorithm for multi-object scenes with context dependencies presented in Section 5.5 was evaluated using the same test images as introduced in Section 6.5.2, i.e., altogether 3240 gray

A-Priori Occurrence Probabilities for Context Scenes										
	Ω_1	Ω_2	Ω_3	Ω_4	Ω_5	Ω_6	Ω_7	Ω_8	Ω_9	Ω_{10}
$\Upsilon_1 = \text{kitchen}$	0.20	0.06	0.06	0.20	0.04	0.06	0.20	0.06	0.06	0.06
$\Upsilon_2 = \text{nursery}$	0.10	0.20	0.04	0.10	0.04	0.20	0.04	0.04	0.20	0.04
$\Upsilon_3 = \text{office}$	0.10	0.04	0.20	0.10	0.04	0.04	0.04	0.20	0.04	0.20

Table 6.9: Learned values of the a-priori occurrence probabilities for all 3D-REAL-ENV object classes in three predefined contexts (6.31).

level multi-object scenes sized 512×512 pixels were used. They were generated based on the 3D-REAL-ENV single-object test images described in detail in Section 6.2.2, and can be divided into three types, i. e., there are 1080 multi-object scenes with homogeneous ($T_{\text{ype}} = \text{hom}$), 1080 multi-object scenes with less heterogeneous ($T_{\text{ype}} = \text{less}$), and 1080 multi-object scenes with more heterogeneous ($T_{\text{ype}} = \text{more}$) background. Additionally, the 3D-REAL-ENV objects (see Figure 6.1) were assigned into three different contexts, namely the kitchen Υ_1, the nursery Υ_2, and the office Υ_3. For each image type ($T_{\text{ype}} \in \{\text{hom}, \text{less}, \text{more}\}$) and each context ($\Upsilon = \{\text{kitchen}, \text{nursery}, \text{office}\}$), 120 one-object, 120 two-object, and 120 three-object scenes were created, whereas the viewpoints were chosen randomly from all 288 3D-REAL-ENV test views (6.14) and are different for all combinations of the test scenes. For example, two-object test scenes with homogeneous background ($T_{\text{ype}} = \text{hom}$) from the kitchen context ($\Upsilon_1 = \text{kitchen}$) represent, in general, different viewpoints as the three-object test scenes with less heterogeneous background ($T_{\text{ype}} = \text{less}$) from the office context ($\Upsilon_3 = \text{office}$). Examples of multi-object test scenes of all types ($T_{\text{ype}} \in \{\text{hom}, \text{less}, \text{more}\}$) and all contexts ($\Upsilon = \{\text{kitchen}, \text{nursery}, \text{office}\}$) are depicted in Figure 6.7. As one can see, even by humans, it is extremely difficult to recognize the 3D-REAL-ENV objects especially in the test images with more heterogeneous background.

Since the number of objects in a multi-object scene is not assumed to be a-priori known, the system determines it first. The robustness of this determination is expressed by the object number determination rate d_r introduced in Section 6.1.2 (6.7). In Table 6.10, the object number determination rates for all types of test images ($T_{\text{ype}} \in \{\text{hom}, \text{less}, \text{more}\}$) are presented; whereas, in contrast to Table 6.7, the context modeling (Section 6.6.1) is taken into consideration in the present section. Comparing the results in Table 6.7 to the object number determination rates in Table 6.10, it can be seen that the using of context modeling increases the robustness of the system in a real world environment.

For object classification, it is assumed that the number of objects in a multi-object scene is

Object Number Determination Rates for Scenes with Context Dependencies		
Homogeneous Background	Less Heterogeneous Background	More Heterogeneous Background
99.9%	88.2%	59.2%

Table 6.10: Object number determination rates for scenes with context dependencies estimated for 1080 test images with homogeneous, 1080 test images with less heterogeneous, and 1080 test images with more heterogeneous background.

Classification and Localization Rates for Scenes with Context Dependencies			
	Homogeneous Background	Less Heterogeneous Background	More Heterogeneous Background
Classification	100%	97.0%	87.5%
Localization	99.7%	81.7%	58.1%

Table 6.11: Classification and localization rates for scenes with context dependencies estimated for 2160 recognition tasks with homogeneous, 2160 recognition tasks with less heterogeneous, and 2160 recognition tasks with more heterogeneous background.

known, while the object localization is made for a-priori known object classes. Furthermore, the context models learned as shown in Section 6.6.1 are taken into account in this section. Considering the configuration of the multi-object test scenes with context dependencies described in Section 6.6.2, there are 1080 three-object, 1080 two-object, and 1080 one-object images. Therefore, altogether 6480 objects have to be classified and localized in these images; whereas 2160 of them are placed on homogeneous, 2160 on less heterogeneous, and 2160 on more heterogeneous background. The classification and localization rates obtained for all multi-object test images (see Figure 6.7) with the recognition algorithm introduced in Section 5.5 can be seen in Table 6.11. Comparing the recognition rates in Table 6.8 to the results in Table 6.11, one can see immediately that the using of the context information brings a significant improvement of the classification and localization in a real world environment.

6.7 Summary

Following the description of the statistical modeling in the training phase (Chapter 4) and the algorithms for object classification and localization (Chapter 5), this chapter evaluated the system with many experiments made in a real world environment.

Section 6.1 listed the hardware and software applied in the experimental phase first (Section 6.1.1). The images used for experiments were acquired with a setup containing turntable and camera arm in a laboratory environment (see Figure 4.2, left). Both the turntable and the robot arm were moved by a Nanotec Step Motor, which was controlled by an Isel CNC-Controller C142-1. Furthermore, a hand-held camera (Sony Digital Camera DSR-PD100A) was used in some cases. Concerning software, the system for appearance-based statistical object classification and localization was developed in object-oriented C++ programming language, whereas some other external software components listed at the end of Section 6.1.1 were also integrated. Finally, Section 6.1 defined coefficients expressing the robustness of the system (Section 6.1.2), namely the classification rate c_r (6.1), the localization rate l_r (6.6), and the object number determination rate d_r (6.7).

Section 6.2 introduced the image database for 3D object recognition in a real world environment (3D-REAL-ENV). This database consists of ten objects ($N_\Omega = 10$) depicted in Figure 6.1. The training images (Section 6.2.1) of these objects were taken on dark background under two different illuminations from 1680 viewpoints (6.10). Thus, there are altogether 33600 training scenes. For testing of the system, three types of test scenes, namely 2880 test images with homogeneous background, 2880 test images with less heterogeneous background, and 2880 test images with more heterogeneous background from 288 different viewpoints were recorded (Section 6.2.2). Illumination in the test scenes is different from the illumination conditions in the training images. The test viewpoints (6.14), in general, are also different from the training points of view (6.10). Additionally, more than 200 different real heterogeneous backgrounds were used for acquisition of the test images.

Results of experiments made for images of single-objects distinguishable with only one view were presented in Section 6.3. The practical realization of the training phase for the 3D-REAL-ENV objects depicted in Figure 6.1 was discussed in Section 6.3.1. Due to the statistical object modeling, a significant compression of the data amount in comparison to the size of training images was achieved (Table 6.2). The classification and localization rates obtained for the 3D-REAL-ENV test images with the recognition algorithm introduced in Section 5.2 were discussed in Section 6.3.2. Due to the color modeling, a significant improvement of the classification and localization rates for test images with more heterogeneous background was obtained (Table 6.3).

The fusion of multiple views realized with the condensation algorithm and described in detail in Section 5.3 was evaluated in Section 6.4. The experiments were performed for the fusion image database with images of single-objects, which are not distinguishable from some points of view (see Figure 6.6). The training phase for the fusion image database was discussed in Section 6.4.1. Section 6.4.2 compared the classification rates obtained using the fusion algorithm (Section 5.3) to the classification results achieved with the simple algorithm for single-object scenes (Section 5.2). Furthermore, it presented the localization rates obtained for this database. Using the fusion algorithm, much better classification rates for the fusion image database were obtained (compare Table 6.4 to Table 6.5).

Experimental results for the algorithm dealing with multi-object scenes without context dependencies, which was theoretical described in Section 5.4, were presented in Section 6.5. The experiments were made on the 3D-REAL-ENV objects depicted in Figure 6.1. The practical realization of the training phase for this algorithm was discussed in Section 6.5.1, while Section 6.5.2 evaluated the object number determination as well as the object classification and localization for this approach. In the testing phase, altogether 3240 gray level multi-object scenes sized 512×512 pixels were used. They were generated based on the 3D-REAL-ENV single-object test images (see Figure 6.7).

Section 6.6 discussed experiments and results for the recognition algorithm dealing with multi-object scenes and considering context dependencies between objects, whereas the same training and test images as in Section 6.5 were used. Section 6.6.1 started with a short revision of the statistical modeling for the 3D-REAL-ENV objects. After that, it presented the practical realization of the context modeling introduced in Section 4.4. The object number determination as well as the object classification and localization found their evaluation in Section 6.6.2. Comparing the results in Table 6.7 to the experimental outcomes in Table 6.10, one can see that the context modeling increases the robustness of the system in terms of object number determination. Moreover, the using of context information improves also the classification and localization rates (compare Table 6.8 to Table 6.11).

Chapter 7

Conclusion

This chapter closes the present work with some conclusions. A summary of the object classification and localization system described in this work is given in Section 7.1. Section 7.2 lists points, in which the approach can be and, for some of them, will be improved in the future.

7.1 Summary

One of the most fundamental problems of computer vision is the recognition of objects in digital images. The term object recognition comprehends both, classification and localization of objects. The task of object classification is to determine the classes of objects occurring in the image $\boldsymbol{f}$ from a set of predefined object classes $\Omega = \{\Omega_1, \Omega_2, \ldots, \Omega_\kappa, \ldots, \Omega_{N_\Omega}\}$. Generally, the number of objects in a scene is unknown. Therefore, it is necessary to find out the number of objects in the image first. In the case of object localization, the recognition system estimates the poses of objects in the image, whereas the object classes are assumed to be a-priori known. The object poses are defined relatively to each other with a 3D translation vector $\boldsymbol{t} = (t_x, t_y, t_z)^{\mathrm{T}}$ and a 3D rotation vector $\boldsymbol{\phi} = (\phi_x, \phi_y, \phi_z)^{\mathrm{T}}$ in a coordinate system with an origin placed in the image center.

There are two main approaches for object recognition, namely shape-based and appearance-based methods. The shape-based algorithms perform a segmentation and use geometric features like lines or corners for object representation [Hor96, Lat00, Lat02, Ker03, Che04]. Common shape-based systems consist usually of three modules, namely the shape representation module, the modeling module, and the classification and localization module [Win94]. Unfortunately, these methods suffer often from segmentation errors. Therefore, many authors, e. g., [Mur95, Pös99, Rei04], prefer a second method, the appearance-based object recognition.

Here, texture is taken into consideration for object description. The object features are computed directly from the pixel values without a previous segmentation step. Most fundamental approaches for appearance-based object classification and localization are template matching [Bru97, Gon01, Pra01], eigenspace approach [Tur91, Leo96, Mog97, Grä03], and Support Vector Machines [Cor95, Vap95].

This work presents an appearance-based approach, which extracts object features directly from image pixels. For the object modeling using gray level images, objects are described by two-dimensional local feature vectors, while in RGB images six-dimensional local feature vectors are computed. The main advantage of the local feature vectors is that a local disturbance affects only the features in a small region around it. In contrast to this, a global feature vector can totally change, if only one pixel in the image varies. The system determines a set of local feature vectors for all training images using the discrete wavelet transformation. The Johnston wavelet and its corresponding scaling function are used for this purpose. In the case of gray level modeling, the first element of the feature vector results from the low-pass filtering in a local neighborhood; while in the second element, information about discontinuities in this local area is stored (high-pass filtering). For color modeling (RGB images), the meaning of the feature vector elements is analogical, whereas the first and second element depend on the red channel, third and fourth on the green channel, and fifth and sixth on the blue channel.

Some feature vectors $\boldsymbol{c}_{\kappa,m}$ describe the object Ω_κ, others belong to the background. In a real world environment, it cannot be assumed that the background in the recognition phase is a-priori known. Therefore, for the statistical object modeling, only feature vectors describing the object should be considered. Since the object takes usually only a part of the image, a tightly enclosing object area O_κ is learned for each object class Ω_κ in the training phase. In the case of 3D objects, the object area changes its location, orientation, and size from image to image. Therefore, it is modeled as a function defined on continuous pose parameter domain $O_\kappa(\boldsymbol{\phi}, \boldsymbol{t})$. In other words, using the object area O_κ, the system is able to determine the set of local feature vectors $\boldsymbol{c}_{\kappa,m}$ describing the object Ω_κ in the whole pose parameter space $(\boldsymbol{\phi}, \boldsymbol{t})$.

In order to handle illumination changes and low-frequency noise, the local feature vectors $\boldsymbol{c}_{\kappa,m}$ are interpreted as random vectors, whereas their elements are assumed to be statistically independent on each other. Since the object feature vectors are also assumed to be statistically independent on the background features, the statistical object modeling is performed separately from the statistical background modeling. For all feature vectors $\boldsymbol{c}_{\kappa,m}$ inside the object area O_κ, corresponding mean vectors $\boldsymbol{\mu}_{\kappa,m}$ and standard deviation vectors $\boldsymbol{\sigma}_{\kappa,m}$ are learned as continuous functions of the object pose parameters $(\boldsymbol{\phi}, \boldsymbol{t})$. Therefore, the object feature vectors are represented by normal density functions $p(\boldsymbol{c}_m|\boldsymbol{\mu}_{\kappa,m}, \boldsymbol{\sigma}_{\kappa,m}, \boldsymbol{\phi}, \boldsymbol{t})$. Additionally, the background feature vectors $\boldsymbol{c}_{\kappa,m}$ lying outside the object area O_κ are modeled with the uniform distribution and represented by constant density functions $p(\boldsymbol{c}_m) = p_b$.

Finally, for all object classes Ω_κ considered in a particular recognition task, statistical object models $\mathcal{M}_\kappa$ are learned in the training phase. The models are regarded as continuous functions defined on the pose parameter domain $\mathcal{M}_\kappa(\boldsymbol{\phi}, \boldsymbol{t})$. This means that the object models $\mathcal{M}_\kappa$ contain the object area O_κ and thereby the set of object feature vectors $\boldsymbol{c}_{\kappa,m}$, the density functions for the object feature vectors $p(\boldsymbol{c}_m|\boldsymbol{\mu}_{\kappa,m}, \boldsymbol{\sigma}_{\kappa,m})$, and the density value for the background features $p(\boldsymbol{c}_m) = p_b$ for all pose parameters $(\boldsymbol{\phi}, \boldsymbol{t})$ in the continuous sense.

It is also possible to learn context dependencies between objects in the training phase. In known environments, the occurrence of some objects is more likely than the occurrence of the others. In this case, the system trains the a-priori occurrence probabilities $p_\iota(\Omega_\kappa)$ for all object classes Ω_κ in all predefined contexts Υ_ι. These discrete density functions are stored in the so-called statistical context models $\mathcal{M}_\iota$.

Since for all object classes Ω_κ regarded in a particular recognition task corresponding object models $\mathcal{M}_\kappa$ and, if considered, for all contexts Υ_ι corresponding context models $\mathcal{M}_\iota$ are learned in the training phase, the system is able to classify and localize objects in images taken from a real world environment. First, a test image is taken, preprocessed, and feature vectors in it are computed. Second, the system starts one of the four recognition algorithms, which are integrated into it. The first algorithm deals with single-object scenes using one image, the second one works also with single-object scenes, but it uses multiple views of the object, the third one is applied for multi-object scenes without context dependencies, and the fourth one considers context dependencies in the multi-object scenes.

The object classification and localization for single-object scenes is solved based on the so-called maximum likelihood estimation. First, local feature vectors are extracted from the preprocessed test image with the wavelet transformation, whereas both gray level and color images can be used. Second, the object area O_κ is determined for all class κ and pose $(\boldsymbol{\phi}_h, \boldsymbol{t}_h)$ hypotheses using the learned object models $\mathcal{M}_\kappa$. Only feature vectors $\boldsymbol{c}_m$ inside the object area are taken into account for evaluation of each hypothesis. Finally, the density values $p(\boldsymbol{c}_m|\boldsymbol{\mu}_{\kappa,m}, \boldsymbol{\sigma}_{\kappa,m})$ for the object feature vectors $\boldsymbol{c}_m$, which are greater than the background density p_b, are multiplied by each other. The result of this multiplication is normalized by a quality measure called geometric criterion and maximized over all class and pose hypotheses in order to find the optimal class $\widehat{\kappa}$ and pose $(\widehat{\boldsymbol{\phi}}, \widehat{\boldsymbol{t}})$ for the test image.

Since some objects cannot be distinguished using only one point of view, the second algorithm for object classification and localization integrated into the system uses a fusion of multiple object views $\langle \boldsymbol{f} \rangle_n = (\boldsymbol{f}_0, \boldsymbol{f}_1, \ldots, \boldsymbol{f}_n)$ for classification, whereas still exactly one resulting object class $\Omega_{\widehat{\kappa}}$ is expected. The object class Ω_κ and pose $(\boldsymbol{\phi}, \boldsymbol{t})$ in the image $\boldsymbol{f}_n$ are called object state $\boldsymbol{q}_n$ in this approach. The task of object recognition with fusion of multiple views is defined by the object state $\boldsymbol{q}_n$ estimation for the given image sequence $\langle \boldsymbol{f} \rangle_n$ and camera movements $\langle \boldsymbol{a} \rangle_{n-1}$ between them. The problem is solved by maximization of the object state density $p(\boldsymbol{q}_n | \boldsymbol{f}_n, \boldsymbol{a}_{n-1}, \ldots, \boldsymbol{a}_0, \boldsymbol{f}_0)$, which is recursively propagated using the condensation algorithm.

The third recognition approach addresses the problem of object classification and localization in multi-object scenes without context modeling. In this case, not only the resulting classes of objects and their poses have to be determined. Since the number of objects $\widehat{\imath}$ in a scene is unknown, it must also be estimated. The a-priori occurrence probability for all object classes Ω_κ is assumed to be equal for this algorithm. Therefore, the maximum likelihood estimation is applied in this case. The first $\widehat{\imath}$ most likely object classes and their best pose estimations are interpreted as the final recognition result.

The last recognition algorithm integrated into the system concerns also the problem of object classification and localization in multi-object scenes. However, it takes into consideration context dependencies between objects, which are statistically modeled in the training phase. Since there is no a-priori knowledge about the context in which a test image was acquired, the algorithm is split into two phases. First, the context $\Upsilon_{\widehat{\iota}}$ of the test image is automatically determined. Second, the approach classifies and localizes objects in the test image $\boldsymbol{f}$ using the statistical context model $\mathcal{M}_{\widehat{\iota}}$ learned for this context $\Upsilon_{\widehat{\iota}}$ in the training phase.

For experiments, an image database for 3D object recognition in a real world environment (3D-REAL-ENV) was generated and is introduced in the scope of the present work. This database consists of ten objects. The training images of these objects were taken on dark background under two different illuminations from 1680 viewpoints. Thus, there are altogether 33600 training scenes. For testing of the system, three types of test scenes, namely 2880 test images with homogeneous background, 2880 test images with less heterogeneous background, and 2880 test images with more heterogeneous background from 288 different viewpoints were acquired. Illumination in the test images is different from the illumination conditions in the training scenes. The test viewpoints, in general, are also different from the training points of view. Additionally, more than 200 different real heterogeneous backgrounds were used for acquiring test images. Due to all these properties, the task of object classification and localization is very difficult for the 3D-REAL-ENV image database.

The classification and localization rates as well as other experimental results obtained with the system described in the present work prove its high performance in a real world environment. Additionally, the two main ideas contributed to the system within the scope of this work, namely the color and context modeling, increase the system robustness significantly. For 3D-REAL-ENV test images with more heterogeneous background, the classification rate achieved with gray level modeling amounts to 54.1%; while applying color modeling for these images, the classification was successful more often (82.3%). The localization algorithm was also improved using color modeling in this really difficult heterogeneous environment, namely from 69.0% (gray level modeling) to 73.6% (color modeling). Furthermore, due to the modeling of context dependencies between objects, much higher classification rates for multi-object scenes were obtained. The classification rate for multi-object scenes with more heterogeneous background achieved without considering context dependencies amounts to 62.9%, while taking into account context modeling it increased to 87.5%.

7.2 Future Work

The approach for appearance-based statistical object classification and localization introduced within the scope of the present work can be and will be improved in some points. Moreover, after some modifications, it can be applied to other tasks of computer vision. The main ideas for future work with the system are following:

- *Partial Object Recognition*
  At the moment, the system deals with whole objects. However, one can imagine to model only a part of an object in order to distinguish it from other objects. In this case a redefinition of the object area O_κ is required. For each class Ω_κ, several object parts $O_{\kappa,i=1,2,\ldots,N_P}$ rather than one object area O_κ concerning the whole object should be learned in the training phase. Evaluating only parts of objects in the recognition phase, the execution time of the algorithm will decrease significantly. Moreover, combining the recognition results for different object parts, an improvement of the recognition rates will be obtained.

- *Video Shot Detection*
  One of the tasks in video mining is the detection of shots in digital videos. Nowadays MPEG-7 standard descriptors [Man02] are often used in order to extract features in video frames, which are then compared to each other by a similarity measure. However, the local feature extraction with the wavelet transformation introduced in this work can be also applied for this purpose. The features can be modeled by density functions, and the

similarity between frames can be determined by a statistical algorithm instead of using a simple distance function. Therefore, the video shot detection problem can also be solved using the probabilistic approach presented in this work.

- *Automatic Image Retrieval*
  Another area where the system introduced in this work can be applied, is automatic image retrieval. An image retrieval system is a computer system for browsing, searching, and retrieving images from a large database of digital images. In order to perform the process automatically without previous annotation step, a robust approach for image classification is required. The task of image classification is different from the object recognition problem. In order to classify images, the system searches for a generic class called concept, like sky, animals, buildings, etc., rather than for one of the predefined object classes. However, it is possible to adapt the statistical approach described in the present work for the image classification and, thereby, for the automatic image retrieval.

- *Texture and Shape for Object Recognition*
  Approaches for object recognition can be generally divided into two categories; appearance-based and shape-based methods. Appearance-based recognition algorithms describe objects based on their texture, while shape-based systems perform a segmentation and use geometric features like lines or corners for object representation. The present work describes an appearance-based approach for object classification and localization, because the shape-based methods seem to suffer from many disadvantages. However, a combination of these two ways for object representation will bring better recognition results. There are objects with the same shape, which are distinguishable only by texture, but one can imagine also objects with the same texture features, which are easy to distinguish by shape.

Additionally to these four areas listed above, the system can be adapted for application to many other tasks within the field of statistical pattern recognition. Further scientific investigations with the system are planned and will take place in the future.

Appendix A

Further Experimental Results

On the following 12 pages tables with classification confusion matrices for the 3D-REAL-ENV image database can be found. Since for test images with homogeneous background ($T_{\text{ype}} =$ hom) the classification rates are close to 100% independent on the distance of training views (see Table 6.3), the following tables present exclusively results for test images with less heterogeneous ($T_{\text{ype}} =$ less) and more heterogeneous ($T_{\text{ype}} =$ more) background. Both gray level and color images are considered. The distance of training views changes from $4.5°$ to $27°$ in 5 steps. The image database for 3D object recognition in a real world environment (3D-REAL-ENV) is described in detail in Section 6.2.

Gray Level Images with Less Heterogeneous Background, Training Distance $4.5°$										
	Ω_1	Ω_2	Ω_3	Ω_4	Ω_5	Ω_6	Ω_7	Ω_8	Ω_9	Ω_{10}
Ω_1 = bscup	99.3	0.0	0.0	0.3	0.3	0.0	0.0	0.0	0.0	0.0
Ω_2 = fireeng	0.0	84.7	2.1	0.0	1.0	0.0	0.0	0.0	12.2	0.0
Ω_3 = greenpunch	0.0	0.0	99.0	0.0	1.0	0.0	0.0	0.0	0.0	0.0
Ω_4 = medcup	0.0	0.0	0.0	100.0	0.0	0.0	0.0	0.0	0.0	0.0
Ω_5 = nizoral	0.0	0.0	1.0	0.0	86.5	0.0	8.0	0.0	4.2	0.3
Ω_6 = perrier	0.0	2.1	0.0	0.0	1.7	90.3	0.0	0.7	5.2	0.0
Ω_7 = ricola	0.0	0.0	0.0	0.0	0.0	0.0	89.6	0.0	10.4	0.0
Ω_8 = stapler	0.0	1.4	0.3	1.0	1.0	0.0	0.0	89.6	5.9	0.7
Ω_9 = truck	0.0	1.4	1.0	0.0	3.1	1.0	0.0	0.0	93.4	0.0
Ω_{10} = whitepunch	0.7	0.0	0.0	0.0	0.3	0.0	0.3	0.0	8.7	89.9

Table A.1: Classification confusion matrix for 2880 3D-REAL-ENV one-object gray level scenes with less heterogeneous background. The distance of training views amounts to $4.5°$. All numbers are given in [%].

Color Images with Less Heterogeneous Background, Training Distance $4.5°$										
	Ω_1	Ω_2	Ω_3	Ω_4	Ω_5	Ω_6	Ω_7	Ω_8	Ω_9	Ω_{10}
Ω_1 = bscup	74.7	0.0	0.0	2.8	12.8	8.0	0.0	0.0	1.7	0.0
Ω_2 = fireeng	0.0	99.3	0.0	0.0	0.7	0.0	0.0	0.0	0.0	0.0
Ω_3 = greenpunch	0.0	0.0	64.9	0.0	20.8	2.4	0.0	0.0	11.8	0.0
Ω_4 = medcup	1.0	0.0	0.0	56.6	39.2	1.4	0.0	0.0	1.7	0.0
Ω_5 = nizoral	0.0	0.0	0.0	0.0	100.0	0.0	0.0	0.0	0.0	0.0
Ω_6 = perrier	0.0	0.0	0.0	0.0	0.7	99.3	0.0	0.0	0.0	0.0
Ω_7 = ricola	0.0	0.0	0.0	0.0	0.0	0.0	100.0	0.0	0.0	0.0
Ω_8 = stapler	0.0	0.0	0.0	0.0	0.0	0.0	0.0	99.7	0.3	0.0
Ω_9 = truck	0.0	0.0	0.0	0.0	1.7	0.0	0.0	0.0	98.3	0.0
Ω_{10} = whitepunch	0.0	0.0	0.0	2.4	9.7	0.3	0.0	0.0	0.3	87.2

Table A.2: Classification confusion matrix for 2880 3D-REAL-ENV one-object color scenes with less heterogeneous background. The distance of training views amounts to $4.5°$. All numbers are given in [%].

Gray Level Images with More Heterogeneous Background, Training Distance 4.5°										
	Ω_1	Ω_2	Ω_3	Ω_4	Ω_5	Ω_6	Ω_7	Ω_8	Ω_9	Ω_{10}
Ω_1 = bscup	100.0	0.0	0.0	0.0	0.0	0.0	0.0	0.0	0.0	0.0
Ω_2 = fireeng	5.9	1.0	9.7	1.4	2.8	0.0	69.1	0.0	3.8	6.2
Ω_3 = greenpunch	0.3	0.7	88.9	0.0	1.0	0.0	7.6	0.0	1.4	0.0
Ω_4 = medcup	0.0	0.0	0.0	100.0	0.0	0.0	0.0	0.0	0.0	0.0
Ω_5 = nizoral	2.8	0.0	0.0	47.9	0.3	0.0	45.5	0.0	0.0	3.5
Ω_6 = perrier	9.7	0.0	2.4	11.5	4.9	10.1	37.2	0.0	0.0	24.3
Ω_7 = ricola	0.0	0.0	0.0	0.0	0.0	0.0	100.0	0.0	0.0	0.0
Ω_8 = stapler	0.0	2.8	4.9	2.4	0.3	0.0	68.8	14.6	1.0	5.2
Ω_9 = truck	5.2	3.8	3.5	13.5	4.2	2.1	28.1	0.0	38.9	0.7
Ω_{10} = whitepunch	1.7	1.0	2.8	1.0	0.0	0.0	4.5	0.0	1.4	87.5

Table A.3: Classification confusion matrix for 2880 3D-REAL-ENV one-object gray level scenes with more heterogeneous background. The distance of training views amounts to 4.5°. All numbers are given in [%].

Color Images with More Heterogeneous Background, Training Distance 4.5°										
	Ω_1	Ω_2	Ω_3	Ω_4	Ω_5	Ω_6	Ω_7	Ω_8	Ω_9	Ω_{10}
Ω_1 = bscup	98.6	0.0	0.0	1.4	0.0	0.0	0.0	0.0	0.0	0.0
Ω_2 = fireeng	1.0	83.3	0.0	7.6	4.2	1.0	0.0	0.0	0.3	2.4
Ω_3 = greenpunch	0.3	0.3	78.1	12.5	5.6	0.0	0.0	0.0	2.4	0.7
Ω_4 = medcup	0.0	0.0	0.0	100.0	0.0	0.0	0.0	0.0	0.0	0.0
Ω_5 = nizoral	7.6	0.0	0.0	2.4	89.9	0.0	0.0	0.0	0.0	0.0
Ω_6 = perrier	12.2	0.0	0.0	27.1	5.2	55.6	0.0	0.0	0.0	0.0
Ω_7 = ricola	0.0	0.0	0.0	0.7	0.0	0.0	99.3	0.0	0.0	0.0
Ω_8 = stapler	2.1	0.0	0.0	5.6	0.3	0.7	0.0	88.9	0.0	2.4
Ω_9 = truck	6.6	1.4	1.4	27.8	13.9	3.5	0.0	5.6	38.5	1.4
Ω_{10} = whitepunch	0.3	0.0	0.3	6.2	1.0	0.0	0.0	0.0	1.7	90.3

Table A.4: Classification confusion matrix for 2880 3D-REAL-ENV one-object color scenes with more heterogeneous background. The distance of training views amounts to 4.5°. All numbers are given in [%].

Gray Level Images with Less Heterogeneous Background, Training Distance 9°										
	Ω_1	Ω_2	Ω_3	Ω_4	Ω_5	Ω_6	Ω_7	Ω_8	Ω_9	Ω_{10}
Ω_1 = bscup	99.7	0.0	0.0	0.0	0.3	0.0	0.0	0.0	0.0	0.0
Ω_2 = fireeng	0.0	83.7	2.8	0.0	1.7	0.0	0.0	0.0	11.8	0.0
Ω_3 = greenpunch	0.0	0.0	99.7	0.0	0.3	0.0	0.0	0.0	0.0	0.0
Ω_4 = medcup	0.0	0.0	0.0	100.0	0.0	0.0	0.0	0.0	0.0	0.0
Ω_5 = nizoral	0.0	0.0	2.1	0.0	85.8	0.0	6.2	0.0	5.6	0.3
Ω_6 = perrier	0.0	1.7	0.0	0.0	4.2	88.2	0.0	0.7	5.2	0.0
Ω_7 = ricola	0.0	0.0	0.0	0.0	0.0	0.0	90.6	0.0	9.4	0.0
Ω_8 = stapler	0.0	1.4	1.0	0.7	2.8	0.0	0.0	89.9	3.8	0.3
Ω_9 = truck	0.0	0.7	0.7	0.0	3.1	0.7	0.0	0.3	94.4	0.0
Ω_{10} = whitepunch	0.3	0.0	0.0	0.0	0.0	0.0	0.3	0.0	6.9	92.4

Table A.5: Classification confusion matrix for 2880 3D-REAL-ENV one-object gray level scenes with less heterogeneous background. The distance of training views amounts to 9°. All numbers are given in [%].

Color Images with Less Heterogeneous Background, Training Distance 9°										
	Ω_1	Ω_2	Ω_3	Ω_4	Ω_5	Ω_6	Ω_7	Ω_8	Ω_9	Ω_{10}
Ω_1 = bscup	73.3	0.0	0.0	3.5	13.5	8.3	0.0	0.0	1.4	0.0
Ω_2 = fireeng	0.0	97.9	0.0	0.0	2.1	0.0	0.0	0.0	0.0	0.0
Ω_3 = greenpunch	0.0	0.3	64.9	0.0	20.5	2.1	0.0	0.0	12.2	0.0
Ω_4 = medcup	0.3	0.0	0.0	61.5	35.4	2.1	0.0	0.0	0.7	0.0
Ω_5 = nizoral	0.0	0.0	0.0	0.0	100.0	0.0	0.0	0.0	0.0	0.0
Ω_6 = perrier	0.0	0.0	0.0	0.0	1.0	99.0	0.0	0.0	0.0	0.0
Ω_7 = ricola	0.0	0.0	0.0	0.0	0.0	0.0	100.0	0.0	0.0	0.0
Ω_8 = stapler	0.0	0.0	0.0	0.0	0.3	0.0	0.0	99.7	0.0	0.0
Ω_9 = truck	0.0	0.0	0.0	0.0	1.4	0.0	0.0	0.0	98.6	0.0
Ω_{10} = whitepunch	0.0	0.0	0.0	2.8	7.3	1.4	0.0	0.0	0.0	88.5

Table A.6: Classification confusion matrix for 2880 3D-REAL-ENV one-object color scenes with less heterogeneous background. The distance of training views amounts to 9°. All numbers are given in [%].

Gray Level Images with More Heterogeneous Background, Training Distance $9°$										
	Ω_1	Ω_2	Ω_3	Ω_4	Ω_5	Ω_6	Ω_7	Ω_8	Ω_9	Ω_{10}
Ω_1 = bscup	100.0	0.0	0.0	0.0	0.0	0.0	0.0	0.0	0.0	0.0
Ω_2 = fireeng	3.5	1.4	14.2	0.7	4.2	0.0	64.2	0.0	4.2	7.6
Ω_3 = greenpunch	0.0	1.0	86.8	0.0	2.1	0.0	9.0	0.0	1.0	0.0
Ω_4 = medcup	0.3	0.0	0.0	99.7	0.0	0.0	0.0	0.0	0.0	0.0
Ω_5 = nizoral	1.0	0.0	0.0	45.8	1.4	0.0	49.0	0.0	0.0	2.8
Ω_6 = perrier	8.7	0.0	3.1	7.6	5.9	11.1	38.2	0.0	0.7	24.7
Ω_7 = ricola	0.0	0.0	0.0	0.0	0.0	0.0	100.0	0.0	0.0	0.0
Ω_8 = stapler	0.0	1.7	3.8	2.4	0.3	0.0	63.9	21.2	1.0	5.6
Ω_9 = truck	4.5	4.2	3.8	11.5	4.2	0.7	26.4	0.0	43.8	1.0
Ω_{10} = whitepunch	1.4	1.0	1.7	0.0	0.7	0.0	4.9	0.0	1.4	88.9

Table A.7: Classification confusion matrix for 2880 3D-REAL-ENV one-object gray level scenes with more heterogeneous background. The distance of training views amounts to $9°$. All numbers are given in [%].

Color Images with More Heterogeneous Background, Training Distance $9°$										
	Ω_1	Ω_2	Ω_3	Ω_4	Ω_5	Ω_6	Ω_7	Ω_8	Ω_9	Ω_{10}
Ω_1 = bscup	96.5	0.0	0.0	3.5	0.0	0.0	0.0	0.0	0.0	0.0
Ω_2 = fireeng	1.4	81.9	0.0	8.7	4.2	1.4	0.0	0.0	0.0	2.4
Ω_3 = greenpunch	0.0	0.0	74.0	15.3	6.6	0.0	0.0	0.0	3.5	0.7
Ω_4 = medcup	0.0	0.0	0.0	100.0	0.0	0.0	0.0	0.0	0.0	0.0
Ω_5 = nizoral	6.9	0.0	0.0	2.8	90.3	0.0	0.0	0.0	0.0	0.0
Ω_6 = perrier	7.3	0.0	0.0	29.5	8.7	54.5	0.0	0.0	0.0	0.0
Ω_7 = ricola	0.0	0.0	0.0	0.3	0.0	0.0	99.7	0.0	0.0	0.0
Ω_8 = stapler	1.4	0.0	0.0	6.9	0.0	0.3	0.0	87.8	0.3	3.1
Ω_9 = truck	5.2	0.3	2.1	28.1	14.2	4.2	0.0	6.2	37.2	2.4
Ω_{10} = whitepunch	0.3	0.0	0.3	7.3	0.3	0.0	0.0	0.0	1.0	90.6

Table A.8: Classification confusion matrix for 2880 3D-REAL-ENV one-object color scenes with more heterogeneous background. The distance of training views amounts to $9°$. All numbers are given in [%].

Gray Level Images with Less Heterogeneous Background, Training Distance $13.5°$										
	Ω_1	Ω_2	Ω_3	Ω_4	Ω_5	Ω_6	Ω_7	Ω_8	Ω_9	Ω_{10}
$\Omega_1 = \mathrm{bscup}$	99.0	0.0	0.0	0.0	0.7	0.0	0.0	0.0	0.3	0.0
$\Omega_2 = \mathrm{fireeng}$	0.0	81.9	2.1	0.0	3.5	0.0	0.0	1.7	10.8	0.0
$\Omega_3 = \mathrm{greenpunch}$	0.0	1.0	95.5	0.0	3.5	0.0	0.0	0.0	0.0	0.0
$\Omega_4 = \mathrm{medcup}$	0.0	0.0	0.0	100.0	0.0	0.0	0.0	0.0	0.0	0.0
$\Omega_5 = \mathrm{nizoral}$	0.0	0.0	1.4	0.0	79.5	0.0	12.8	0.0	6.2	0.0
$\Omega_6 = \mathrm{perrier}$	0.0	2.1	0.3	0.0	6.6	82.6	0.0	1.0	7.3	0.0
$\Omega_7 = \mathrm{ricola}$	0.0	0.0	0.0	0.0	0.0	0.0	91.3	0.0	8.7	0.0
$\Omega_8 = \mathrm{stapler}$	0.0	0.3	1.4	2.1	2.1	0.0	0.3	87.2	5.9	0.7
$\Omega_9 = \mathrm{truck}$	0.0	1.0	0.3	0.0	5.9	0.0	0.0	0.3	92.4	0.0
$\Omega_{10} = \mathrm{whitepunch}$	0.3	0.0	0.7	0.0	2.4	0.0	0.3	0.0	8.7	87.5

Table A.9: Classification confusion matrix for 2880 3D-REAL-ENV one-object gray level scenes with less heterogeneous background. The distance of training views amounts to $13.5°$. All numbers are given in [%].

Color Images with Less Heterogeneous Background, Training Distance $13.5°$										
	Ω_1	Ω_2	Ω_3	Ω_4	Ω_5	Ω_6	Ω_7	Ω_8	Ω_9	Ω_{10}
$\Omega_1 = \mathrm{bscup}$	68.1	0.0	0.0	3.5	18.1	10.1	0.0	0.0	0.3	0.0
$\Omega_2 = \mathrm{fireeng}$	0.0	95.5	0.0	0.0	4.2	0.0	0.0	0.0	0.3	0.0
$\Omega_3 = \mathrm{greenpunch}$	0.0	0.0	45.5	0.0	41.0	3.8	0.0	0.0	9.7	0.0
$\Omega_4 = \mathrm{medcup}$	1.7	0.0	0.0	47.2	49.3	1.4	0.0	0.0	0.3	0.0
$\Omega_5 = \mathrm{nizoral}$	0.0	0.0	0.0	0.0	99.0	0.0	0.0	0.0	1.0	0.0
$\Omega_6 = \mathrm{perrier}$	0.0	0.3	0.0	0.0	1.0	98.6	0.0	0.0	0.0	0.0
$\Omega_7 = \mathrm{ricola}$	0.0	0.0	0.0	0.0	0.0	0.0	99.3	0.0	0.7	0.0
$\Omega_8 = \mathrm{stapler}$	0.0	0.0	0.0	0.0	2.8	1.0	0.0	95.8	0.3	0.0
$\Omega_9 = \mathrm{truck}$	0.0	0.0	0.0	0.0	7.3	0.0	0.0	0.0	92.7	0.0
$\Omega_{10} = \mathrm{whitepunch}$	0.0	0.0	0.7	0.7	11.1	1.0	0.0	0.0	1.0	85.4

Table A.10: Classification confusion matrix for 2880 3D-REAL-ENV one-object color scenes with less heterogeneous background. The distance of training views amounts to $13.5°$. All numbers are given in [%].

Gray Level Images with More Heterogeneous Background, Training Distance 13.5°

	Ω_1	Ω_2	Ω_3	Ω_4	Ω_5	Ω_6	Ω_7	Ω_8	Ω_9	Ω_{10}
Ω_1 = bscup	99.3	0.0	0.0	0.7	0.0	0.0	0.0	0.0	0.0	0.0
Ω_2 = fireeng	1.4	6.6	8.0	1.7	3.1	0.0	63.5	0.0	7.3	8.3
Ω_3 = greenpunch	0.0	1.7	88.5	0.0	2.4	0.0	6.6	0.0	0.3	0.3
Ω_4 = medcup	0.0	0.0	0.0	100.0	0.0	0.0	0.0	0.0	0.0	0.0
Ω_5 = nizoral	0.7	0.0	0.0	48.3	0.0	0.0	44.4	0.0	0.0	6.6
Ω_6 = perrier	4.9	0.0	1.7	12.5	5.6	9.7	41.7	0.0	0.7	23.3
Ω_7 = ricola	0.0	0.0	0.0	0.0	0.0	0.0	100.0	0.0	0.0	0.0
Ω_8 = stapler	0.3	2.4	8.0	2.4	1.0	0.0	68.4	14.2	1.0	2.1
Ω_9 = truck	1.7	2.1	1.4	10.1	9.0	0.0	18.8	1.0	55.6	0.3
Ω_{10} = whitepunch	0.7	0.7	1.4	0.3	1.7	0.0	4.9	0.0	2.1	88.2

Table A.11: Classification confusion matrix for 2880 3D-REAL-ENV one-object gray level scenes with more heterogeneous background. The distance of training views amounts to 13.5°. All numbers are given in [%].

Color Images with More Heterogeneous Background, Training Distance 13.5°

	Ω_1	Ω_2	Ω_3	Ω_4	Ω_5	Ω_6	Ω_7	Ω_8	Ω_9	Ω_{10}
Ω_1 = bscup	99.7	0.0	0.0	0.3	0.0	0.0	0.0	0.0	0.0	0.0
Ω_2 = fireeng	1.0	79.2	0.0	2.8	7.6	0.7	0.0	0.0	0.3	8.3
Ω_3 = greenpunch	1.0	0.0	73.3	13.2	10.8	0.3	0.0	0.0	0.3	1.0
Ω_4 = medcup	1.0	0.0	0.0	99.0	0.0	0.0	0.0	0.0	0.0	0.0
Ω_5 = nizoral	7.6	0.0	0.0	5.9	84.7	0.0	0.0	0.0	0.0	1.7
Ω_6 = perrier	4.5	0.0	0.0	23.3	9.4	60.8	0.0	0.0	0.0	2.1
Ω_7 = ricola	0.0	0.0	0.0	1.0	2.1	0.3	96.2	0.0	0.0	0.3
Ω_8 = stapler	1.0	0.0	0.0	5.6	2.1	0.3	0.0	83.7	0.3	6.9
Ω_9 = truck	5.6	0.0	1.7	24.3	23.6	1.7	0.0	3.5	31.2	8.3
Ω_{10} = whitepunch	0.0	0.0	0.3	1.7	1.4	0.3	0.0	0.0	0.7	95.5

Table A.12: Classification confusion matrix for 2880 3D-REAL-ENV one-object color scenes with more heterogeneous background. The distance of training views amounts to 13.5°. All numbers are given in [%].

Gray Level Images with Less Heterogeneous Background, Training Distance $18°$										
	Ω_1	Ω_2	Ω_3	Ω_4	Ω_5	Ω_6	Ω_7	Ω_8	Ω_9	Ω_{10}
Ω_1 = bscup	97.2	0.0	0.0	0.0	1.4	0.3	0.3	0.0	0.7	0.0
Ω_2 = fireeng	0.0	82.6	2.8	0.0	2.1	1.4	0.0	1.0	10.1	0.0
Ω_3 = greenpunch	0.0	1.7	96.5	0.0	0.3	0.0	0.0	1.4	0.0	0.0
Ω_4 = medcup	0.0	0.0	0.0	100.0	0.0	0.0	0.0	0.0	0.0	0.0
Ω_5 = nizoral	0.0	0.0	0.7	0.0	84.4	0.0	8.3	0.0	6.2	0.3
Ω_6 = perrier	0.0	1.0	0.0	0.0	14.6	76.4	0.0	1.4	6.6	0.0
Ω_7 = ricola	0.0	0.0	0.0	0.0	0.0	0.0	99.3	0.0	0.7	0.0
Ω_8 = stapler	0.0	1.0	1.4	0.3	3.5	0.0	0.3	76.7	16.3	0.3
Ω_9 = truck	0.0	2.4	0.3	0.0	7.6	2.1	0.0	1.0	86.5	0.0
Ω_{10} = whitepunch	0.3	0.0	0.0	0.0	2.4	0.0	1.0	0.0	3.8	92.4

Table A.13: Classification confusion matrix for 2880 3D-REAL-ENV one-object gray level scenes with less heterogeneous background. The distance of training views amounts to 18°. All numbers are given in [%].

Color Images with Less Heterogeneous Background, Training Distance $18°$										
	Ω_1	Ω_2	Ω_3	Ω_4	Ω_5	Ω_6	Ω_7	Ω_8	Ω_9	Ω_{10}
Ω_1 = bscup	73.6	0.0	0.0	1.4	20.1	3.8	0.0	0.0	1.0	0.0
Ω_2 = fireeng	0.0	97.2	0.0	0.0	2.8	0.0	0.0	0.0	0.0	0.0
Ω_3 = greenpunch	0.0	0.7	28.5	0.0	57.3	1.4	0.0	0.0	12.2	0.0
Ω_4 = medcup	1.0	0.0	0.0	52.8	44.4	1.7	0.0	0.0	0.0	0.0
Ω_5 = nizoral	0.0	0.0	0.0	0.0	99.0	0.0	0.0	0.0	1.0	0.0
Ω_6 = perrier	0.0	0.3	0.0	0.0	2.4	97.2	0.0	0.0	0.0	0.0
Ω_7 = ricola	0.0	0.0	0.0	0.0	0.0	0.0	100.0	0.0	0.0	0.0
Ω_8 = stapler	0.0	0.3	0.0	0.0	5.6	0.0	0.0	87.8	6.2	0.0
Ω_9 = truck	0.0	0.0	0.0	0.0	6.9	0.0	0.0	0.0	93.1	0.0
Ω_{10} = whitepunch	0.3	0.3	0.0	9.0	12.5	0.7	0.0	0.0	0.7	76.4

Table A.14: Classification confusion matrix for 2880 3D-REAL-ENV one-object color scenes with less heterogeneous background. The distance of training views amounts to 18°. All numbers are given in [%].

Gray Level Images with More Heterogeneous Background, Training Distance 18°										
	Ω_1	Ω_2	Ω_3	Ω_4	Ω_5	Ω_6	Ω_7	Ω_8	Ω_9	Ω_{10}
Ω_1 = bscup	99.0	0.0	0.0	1.0	0.0	0.0	0.0	0.0	0.0	0.0
Ω_2 = fireeng	0.3	3.1	18.4	0.0	11.5	0.0	53.8	0.0	8.7	4.2
Ω_3 = greenpunch	0.0	1.7	90.3	0.0	1.4	0.0	4.9	0.0	0.7	1.0
Ω_4 = medcup	1.7	0.0	0.0	98.3	0.0	0.0	0.0	0.0	0.0	0.0
Ω_5 = nizoral	0.0	0.0	0.0	44.4	0.0	0.0	47.9	0.0	0.0	7.6
Ω_6 = perrier	3.1	0.0	3.5	8.7	18.4	8.7	42.0	0.0	1.7	13.9
Ω_7 = ricola	0.0	0.0	0.0	0.0	0.0	0.0	100.0	0.0	0.0	0.0
Ω_8 = stapler	0.0	1.4	5.6	1.7	4.2	0.0	68.1	16.7	1.7	0.7
Ω_9 = truck	1.7	5.6	1.0	9.0	9.7	1.0	19.8	1.4	50.7	0.0
Ω_{10} = whitepunch	0.3	1.4	1.7	0.7	3.1	0.0	6.2	0.0	1.7	84.7

Table A.15: Classification confusion matrix for 2880 3D-REAL-ENV one-object gray level scenes with more heterogeneous background. The distance of training views amounts to 18°. All numbers are given in [%].

Color Images with More Heterogeneous Background, Training Distance 18°										
	Ω_1	Ω_2	Ω_3	Ω_4	Ω_5	Ω_6	Ω_7	Ω_8	Ω_9	Ω_{10}
Ω_1 = bscup	99.3	0.0	0.0	0.7	0.0	0.0	0.0	0.0	0.0	0.0
Ω_2 = fireeng	0.7	67.4	0.0	13.5	12.8	1.7	0.0	0.0	0.7	3.1
Ω_3 = greenpunch	0.0	0.0	58.0	24.3	15.3	1.0	0.0	0.0	1.4	0.0
Ω_4 = medcup	0.3	0.0	0.0	99.7	0.0	0.0	0.0	0.0	0.0	0.0
Ω_5 = nizoral	8.0	0.0	0.0	13.5	78.5	0.0	0.0	0.0	0.0	0.0
Ω_6 = perrier	7.6	0.0	0.0	28.8	33.3	30.2	0.0	0.0	0.0	0.0
Ω_7 = ricola	0.0	0.0	0.0	0.3	2.8	0.0	96.9	0.0	0.0	0.0
Ω_8 = stapler	3.8	0.0	0.0	7.3	3.8	20.5	0.0	58.0	3.8	2.8
Ω_9 = truck	5.2	1.0	0.0	24.3	34.4	3.5	0.0	3.1	22.2	6.2
Ω_{10} = whitepunch	0.0	0.0	0.0	11.8	11.8	0.0	0.0	0.0	0.0	76.4

Table A.16: Classification confusion matrix for 2880 3D-REAL-ENV one-object color scenes with more heterogeneous background. The distance of training views amounts to 18°. All numbers are given in [%].

Gray Level Images with Less Heterogeneous Background, Training Distance $22.5°$										
	Ω_1	Ω_2	Ω_3	Ω_4	Ω_5	Ω_6	Ω_7	Ω_8	Ω_9	Ω_{10}
Ω_1 = bscup	91.3	0.0	0.0	1.7	0.7	0.0	0.0	0.0	6.2	0.0
Ω_2 = fireeng	0.0	83.3	2.1	0.0	0.7	0.3	0.0	3.8	9.7	0.0
Ω_3 = greenpunch	0.0	1.4	90.6	0.0	0.3	0.0	0.0	6.9	0.7	0.0
Ω_4 = medcup	0.0	0.0	0.0	99.7	0.3	0.0	0.0	0.0	0.0	0.0
Ω_5 = nizoral	0.0	0.0	0.3	0.0	88.2	0.3	4.2	0.7	4.2	2.1
Ω_6 = perrier	0.0	1.0	0.0	0.0	14.6	65.3	0.0	3.1	16.0	0.0
Ω_7 = ricola	0.0	0.0	0.0	0.0	0.3	0.0	99.3	0.0	0.3	0.0
Ω_8 = stapler	0.0	2.1	0.7	0.7	3.8	0.0	0.0	67.0	25.0	0.7
Ω_9 = truck	0.0	1.4	0.0	0.0	2.8	0.7	0.0	4.5	90.6	0.0
Ω_{10} = whitepunch	0.3	0.0	0.0	0.3	5.6	0.3	2.1	0.0	6.2	85.1

Table A.17: Classification confusion matrix for 2880 3D-REAL-ENV one-object gray level scenes with less heterogeneous background. The distance of training views amounts to 22.5°. All numbers are given in [%].

Color Images with Less Heterogeneous Background, Training Distance $22.5°$										
	Ω_1	Ω_2	Ω_3	Ω_4	Ω_5	Ω_6	Ω_7	Ω_8	Ω_9	Ω_{10}
Ω_1 = bscup	72.9	1.0	0.0	1.0	17.7	5.6	0.0	0.0	1.7	0.0
Ω_2 = fireeng	0.0	97.2	0.0	0.0	2.8	0.0	0.0	0.0	0.0	0.0
Ω_3 = greenpunch	0.0	9.4	18.1	0.0	59.7	4.9	0.0	0.0	8.0	0.0
Ω_4 = medcup	6.2	0.7	0.0	47.6	41.3	3.5	0.0	0.0	0.7	0.0
Ω_5 = nizoral	0.0	0.0	0.0	0.0	99.0	0.7	0.0	0.0	0.3	0.0
Ω_6 = perrier	0.0	0.7	0.0	0.0	5.6	93.8	0.0	0.0	0.0	0.0
Ω_7 = ricola	0.0	0.3	0.0	0.0	0.3	0.0	99.3	0.0	0.0	0.0
Ω_8 = stapler	0.0	0.7	0.0	0.0	19.4	2.4	0.0	75.7	1.7	0.0
Ω_9 = truck	0.0	0.0	0.0	0.0	12.5	0.3	0.0	0.0	87.2	0.0
Ω_{10} = whitepunch	5.9	0.3	0.0	7.6	26.0	1.7	0.0	0.0	0.7	57.6

Table A.18: Classification confusion matrix for 2880 3D-REAL-ENV one-object color scenes with less heterogeneous background. The distance of training views amounts to 22.5°. All numbers are given in [%].

Gray Level Images with More Heterogeneous Background, Training Distance $22.5°$										
	Ω_1	Ω_2	Ω_3	Ω_4	Ω_5	Ω_6	Ω_7	Ω_8	Ω_9	Ω_{10}
Ω_1 = bscup	96.5	0.0	0.0	2.8	0.7	0.0	0.0	0.0	0.0	0.0
Ω_2 = fireeng	0.0	4.2	11.1	0.3	12.8	0.0	58.3	0.0	11.1	2.1
Ω_3 = greenpunch	0.0	2.8	83.3	0.0	5.2	0.0	6.2	0.3	2.1	0.0
Ω_4 = medcup	2.8	0.0	0.0	97.2	0.0	0.0	0.0	0.0	0.0	0.0
Ω_5 = nizoral	0.7	0.0	0.7	33.7	0.0	0.0	59.4	0.0	0.0	5.6
Ω_6 = perrier	1.0	1.7	2.1	7.6	18.8	10.8	47.2	0.0	2.4	8.3
Ω_7 = ricola	0.0	0.0	0.0	0.0	0.0	0.0	100.0	0.0	0.0	0.0
Ω_8 = stapler	0.0	8.3	4.5	0.7	1.7	0.3	64.6	11.8	8.0	0.0
Ω_9 = truck	0.0	5.2	0.0	8.0	7.3	0.0	17.0	7.3	54.9	0.3
Ω_{10} = whitepunch	1.7	3.1	3.5	0.0	9.7	0.3	9.4	0.0	2.4	69.8

Table A.19: Classification confusion matrix for 2880 3D-REAL-ENV one-object gray level scenes with more heterogeneous background. The distance of training views amounts to $22.5°$. All numbers are given in [%].

Color Images with More Heterogeneous Background, Training Distance $22.5°$										
	Ω_1	Ω_2	Ω_3	Ω_4	Ω_5	Ω_6	Ω_7	Ω_8	Ω_9	Ω_{10}
Ω_1 = bscup	97.6	0.0	0.0	1.7	0.7	0.0	0.0	0.0	0.0	0.0
Ω_2 = fireeng	2.8	48.6	0.0	4.2	29.5	7.3	0.0	0.0	0.3	7.3
Ω_3 = greenpunch	3.5	0.0	51.7	22.9	20.1	1.0	0.0	0.0	0.3	0.3
Ω_4 = medcup	4.5	0.0	0.0	95.5	0.0	0.0	0.0	0.0	0.0	0.0
Ω_5 = nizoral	17.0	0.0	0.0	7.3	75.3	0.3	0.0	0.0	0.0	0.0
Ω_6 = perrier	6.6	0.3	0.0	16.7	52.4	24.0	0.0	0.0	0.0	0.0
Ω_7 = ricola	0.0	0.0	0.0	0.0	5.2	0.0	94.8	0.0	0.0	0.0
Ω_8 = stapler	4.9	2.4	0.3	4.2	23.6	31.9	0.0	27.1	0.7	4.9
Ω_9 = truck	3.8	16.3	0.7	28.8	25.7	4.5	0.0	4.2	13.9	2.1
Ω_{10} = whitepunch	3.1	0.0	0.0	8.3	24.7	0.3	0.0	0.0	0.0	63.5

Table A.20: Classification confusion matrix for 2880 3D-REAL-ENV one-object color scenes with more heterogeneous background. The distance of training views amounts to $22.5°$. All numbers are given in [%].

Gray Level Images with Less Heterogeneous Background, Training Distance 27°										
	Ω_1	Ω_2	Ω_3	Ω_4	Ω_5	Ω_6	Ω_7	Ω_8	Ω_9	Ω_{10}
Ω_1 = bscup	64.2	0.0	0.0	0.0	8.0	0.0	0.0	0.3	27.4	0.0
Ω_2 = fireeng	0.0	67.7	0.0	0.0	6.2	0.0	0.0	9.0	17.0	0.0
Ω_3 = greenpunch	0.0	1.4	62.2	0.0	2.4	0.0	0.0	17.7	16.3	0.0
Ω_4 = medcup	1.4	0.0	0.0	92.4	1.7	0.0	0.0	1.4	2.1	1.0
Ω_5 = nizoral	0.0	0.0	0.0	0.0	95.1	0.0	0.0	1.4	3.5	0.0
Ω_6 = perrier	0.0	2.8	0.0	0.0	16.3	40.3	0.0	10.8	29.9	0.0
Ω_7 = ricola	0.0	0.0	0.0	0.0	9.0	0.0	58.0	1.4	31.6	0.0
Ω_8 = stapler	0.0	1.0	0.0	0.0	4.9	0.0	0.0	67.7	26.4	0.0
Ω_9 = truck	0.0	0.0	0.0	0.0	3.5	0.0	0.0	6.6	89.9	0.0
Ω_{10} = whitepunch	0.0	0.7	0.0	0.0	15.6	0.3	0.0	0.7	25.7	56.9

Table A.21: Classification confusion matrix for 2880 3D-REAL-ENV one-object gray level scenes with less heterogeneous background. The distance of training views amounts to 27°. All numbers are given in [%].

Color Images with Less Heterogeneous Background, Training Distance 27°										
	Ω_1	Ω_2	Ω_3	Ω_4	Ω_5	Ω_6	Ω_7	Ω_8	Ω_9	Ω_{10}
Ω_1 = bscup	24.0	1.0	0.0	11.5	59.7	1.0	0.0	0.0	2.8	0.0
Ω_2 = fireeng	0.0	92.0	0.0	0.0	7.3	0.0	0.0	0.0	0.7	0.0
Ω_3 = greenpunch	0.0	1.0	0.0	0.0	86.5	1.4	0.0	0.0	11.1	0.0
Ω_4 = medcup	1.0	0.3	0.0	30.2	67.7	0.0	0.0	0.0	0.7	0.0
Ω_5 = nizoral	0.0	0.0	0.0	0.0	99.3	0.0	0.0	0.0	0.7	0.0
Ω_6 = perrier	0.0	1.4	0.0	0.0	44.4	52.8	0.0	0.0	1.4	0.0
Ω_7 = ricola	0.0	1.4	0.0	0.0	29.9	0.0	61.1	0.0	7.6	0.0
Ω_8 = stapler	0.0	0.3	0.0	0.0	31.9	0.3	0.0	64.9	2.4	0.0
Ω_9 = truck	0.0	0.0	0.0	0.0	20.5	0.0	0.0	0.0	79.5	0.0
Ω_{10} = whitepunch	1.4	0.0	0.0	4.2	56.2	4.2	0.0	0.0	1.7	32.3

Table A.22: Classification confusion matrix for 2880 3D-REAL-ENV one-object color scenes with less heterogeneous background. The distance of training views amounts to 27°. All numbers are given in [%].

Gray Level Images with More Heterogeneous Background, Training Distance 27°										
	Ω_1	Ω_2	Ω_3	Ω_4	Ω_5	Ω_6	Ω_7	Ω_8	Ω_9	Ω_{10}
Ω_1 = bscup	79.9	0.0	0.0	0.0	18.1	0.3	0.0	0.0	1.7	0.0
Ω_2 = fireeng	0.0	0.3	0.3	0.0	63.2	0.7	0.7	0.0	34.4	0.3
Ω_3 = greenpunch	0.0	2.4	55.6	0.0	27.1	0.0	0.0	0.3	14.6	0.0
Ω_4 = medcup	7.6	0.0	0.3	84.7	5.6	0.0	0.0	0.0	1.7	0.0
Ω_5 = nizoral	0.0	0.0	0.3	0.3	93.1	0.0	1.0	0.0	0.3	4.9
Ω_6 = perrier	0.0	0.0	0.0	0.0	63.2	10.8	1.7	0.0	22.2	2.1
Ω_7 = ricola	0.0	0.0	0.0	0.0	24.7	0.0	71.2	0.0	4.2	0.0
Ω_8 = stapler	0.0	3.1	1.0	0.0	26.0	0.0	1.7	29.9	38.2	0.0
Ω_9 = truck	0.0	4.5	0.0	0.0	21.9	0.0	0.0	4.2	67.7	1.7
Ω_{10} = whitepunch	0.0	5.2	0.0	0.0	31.6	0.0	0.0	0.3	11.8	51.0

Table A.23: Classification confusion matrix for 2880 3D-REAL-ENV one-object gray level scenes with more heterogeneous background. The distance of training views amounts to 27°. All numbers are given in [%].

Color Images with More Heterogeneous Background, Training Distance 27°										
	Ω_1	Ω_2	Ω_3	Ω_4	Ω_5	Ω_6	Ω_7	Ω_8	Ω_9	Ω_{10}
Ω_1 = bscup	91.7	0.0	0.0	2.1	6.2	0.0	0.0	0.0	0.0	0.0
Ω_2 = fireeng	0.0	31.6	0.0	0.3	63.5	2.8	0.0	0.3	0.7	0.7
Ω_3 = greenpunch	1.7	0.0	26.0	12.5	57.3	0.0	0.0	0.0	1.0	1.4
Ω_4 = medcup	2.1	0.0	0.0	93.8	4.2	0.0	0.0	0.0	0.0	0.0
Ω_5 = nizoral	3.8	0.0	0.0	0.0	95.8	0.3	0.0	0.0	0.0	0.0
Ω_6 = perrier	0.3	0.0	0.0	6.9	71.5	17.7	0.0	0.0	0.0	3.5
Ω_7 = ricola	0.3	0.3	0.0	0.3	41.3	0.0	57.3	0.0	0.3	0.0
Ω_8 = stapler	2.4	0.3	0.7	1.7	66.7	0.0	0.0	21.9	2.8	3.5
Ω_9 = truck	9.7	0.0	0.0	20.1	47.9	0.3	0.0	2.4	18.8	0.7
Ω_{10} = whitepunch	0.3	0.0	0.0	3.5	44.8	2.4	0.0	0.0	1.4	47.6

Table A.24: Classification confusion matrix for 2880 3D-REAL-ENV one-object color scenes with more heterogeneous background. The distance of training views amounts to 27°. All numbers are given in [%].

Appendix B

German Version

B.1 Titel

Erscheinungsbasierte, statistische Objekterkennung
mit Farb- und Kontextmodellierung

B.2 Inhaltsverzeichnis

B.3 Einleitung

Jede zielgerichtete menschliche Aktivität erfordert *Perzeption*. Dabei versteht man unter perzeptivem Verhalten das Bemerken, Auswerten und Interpretieren von Sinneseindrücken. Die Perzeption erlaubt dem Menschen, Informationen über seine Umgebung zu erlangen, auf diese zu reagieren, und die Umgebung zu beeinflussen. Es gibt keinen prinzipiellen Grund, weshalb die Perzeption durch ein anderes Wesen oder einen anderen Gegenstand, z. B. einen digitalen Rechner, nicht simuliert werden könnte [Nie90]. Dabei ist das Ziel dieser Simulation nicht die exakte Modellierung der Gehirnaktivitäten eines Menschen, aber vielmehr die Erlangung von ähnlichen perzeptiven Ergebnissen. Das Forschungsgebiet, das sich mit den mathematischen und technischen Aspekten der Perzeption beschäftigt, wird *Mustererkennung* genannt. Das *Sehen*, als eine der wichtigsten perzeptiven Fähigkeiten, liefert visuelle Eindrücke. Die Bearbeitung dieser visuellen Information ist die Aufgabe von *Bildanalyse*, deren Hauptproblem in der Erkennung, Bewertung und Interpretation von bekannten Mustern oder Objekten in Bildern liegt.

B.3.1 Grundkonzept der Objekterkennung

Eines der grundlegendsten Probleme vom *Rechnersehen* ist die Erkennung von Objekten in digitalen Bildern. Dabei versteckt sich hinter dem Begriff *Objekterkennung* sowohl die *Klassifikation* als auch die *Lokalisation* von Objekten.

Um die Aufgabe der Objektklassifikation zu lösen, muss das Objekterkennungssystem die Antwort auf folgende Frage liefern:

Welche Objekte befinden sich im Bild?

Das System ermittelt die Klassen von Objekten, die sich im Bild befinden, wobei die Menge aller möglichen Objektklassen bekannt ist. Die Anzahl der Objekte in einer Szene ist aber im Allgemeinen unbekannt. Zusätzlich kann jedes Objekt mehrmals in einem Bild vorkommen (siehe Abbildung B.2). Deshalb muss die Anzahl der Objekte im Bild zuerst bestimmt werden.

Im Falle der Objektlokalisation liefert das System die Antwort auf folgende Frage:

In welchen Positionen befinden sich bekannte Objekte im Bild?

Unter der Objektposition wird sowohl die Lage als auch die Orientierung eines Objektes verstanden. Sie wird mit einem Translationsvektor $\boldsymbol{t} = (t_x, t_y, t_z)^{\mathrm{T}}$ und drei Rotationswinkeln (ϕ_x, ϕ_y und ϕ_z) um die drei Achsen des kartesischen Koordinatensystems definiert. In der vorliegen-

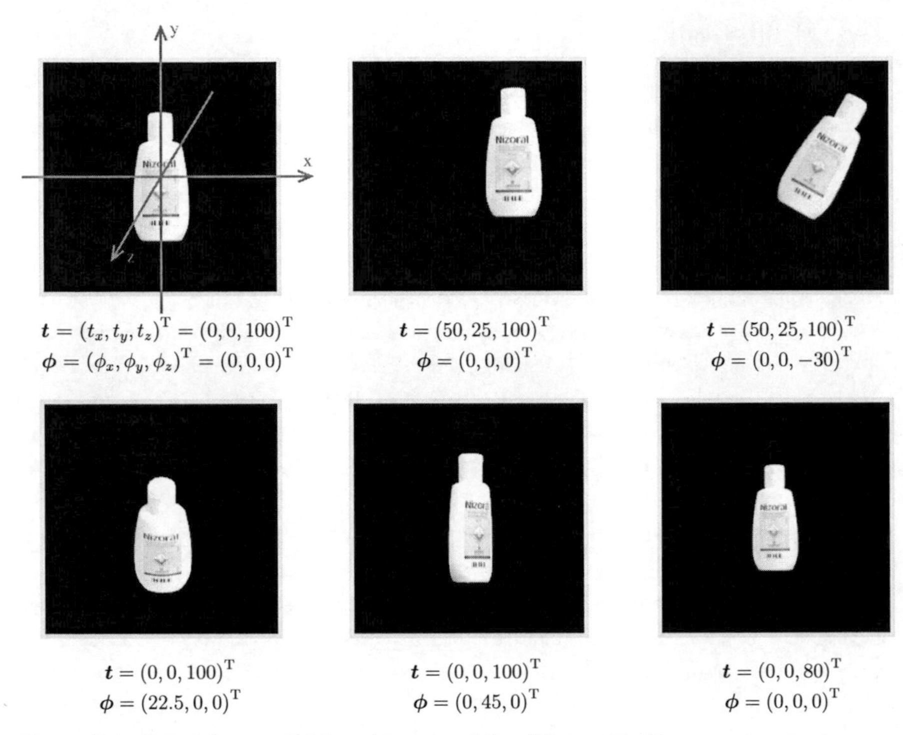

Figure B.1: Beispiele von Objektpositionen und ihre Werte. Die Komponenten des internen Translationsvektors $\boldsymbol{t}_{\text{int}} = (t_x, t_y)^{\mathrm{T}}$ sind in Pixel und die Komponenten des Rotationsvektors in Grad [°] angegeben. Die externe Translation (Skalierung) $t_{\text{ext}} = t_z$ ist in Prozent [%] zur Größe des Referenzbildes (links oben) dargestellt.

den Arbeit werden diese drei Rotationswinkel zu einem Rotationsvektor[1] $\boldsymbol{\phi} = (\phi_x, \phi_y, \phi_z)^{\mathrm{T}}$ zusammengefasst. Der Ursprung des kartesischen Koordinatensystems deckt sich mit dem Mittelpunkt des Bildes, die x- und y-Achsen liegen in in der Bildebene, während die z-Achse senkrecht zu der Bildebene steht. Die Definition der Objektposition ist in der Abbildung B.1 dargestellt. Wenn ein Objekt ausschließlich mit internen Translationsparametern $\boldsymbol{t}_{\text{int}} = (t_x, t_y)^{\mathrm{T}}$ und interner Rotation $\phi_{\text{int}} = \phi_z$ transformiert wird, dann bleibt seine Größe und sein Aussehen konstant. Dieser zweidimensionale Fall ist in der ersten Reihe der Abbildung B.1 illustriert.

[1] Das Ziel der Objektlokalisation ist die Bestimmung der Positionsparameter und keine Transformation von 3D Punkten. Deshalb wird in dieser Arbeit der Begriff Rotationsvektor anstatt von Rotationsmatrix verwendet.

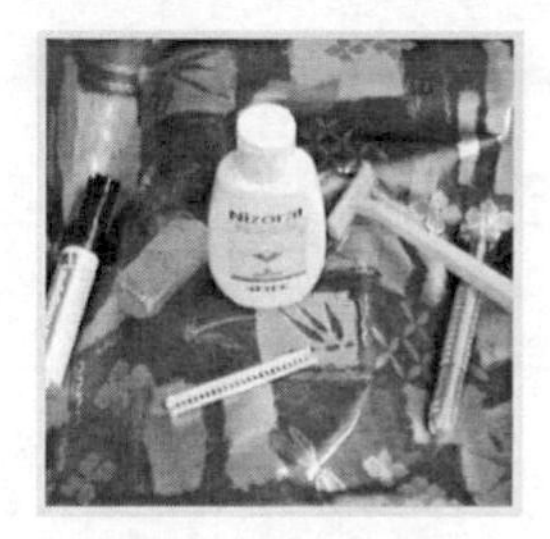

Figure B.2: Komplexe Szenen mit heterogenem Hintergrund, in denen die Objekte sehr schwierig vom Hintergrund zu unterscheiden sind. Die Objekte sind teilweise verdeckt und ihre Anzahl ist unbekannt.

Eine interessantere Situation tritt auf, wenn das Objekt mit externen Transformationsparametern $t_{\text{ext}} = t_z$ und $\boldsymbol{\phi}_{\text{ext}} = (\phi_x, \phi_y)^{\text{T}}$ transformiert wird. In der zweiten Reihe der Abbildung B.1 verändert sich nicht nur die Größe, sondern auch das Aussehen des Objektes. Das Ziel der Objektlokalisation besteht darin, in einer realen Umgebung sowohl die internen als auch die externen Objektpositionsparameter zu bestimmen.

Das Problem der Objekterkennung, d. h. der Klassifikation und Lokalisation von Objekten, kann im Allgemeinen mit folgender Formulierung beschrieben werden:

Welche Objekte befinden sich im Bild und in welchen Positionen?

Die Verfahren zur Objekterkennung lassen sich in zwei Gruppen unterteilen. Die *formbasierten* Ansätze führen einen Segmentierungsschritt durch und verwenden geometrische Merkmale wie Linien und Ecken, um Objekte zu beschreiben [Hor96, Ker03, Che04, Lat05]. Leider werden diese Methoden oft durch Segmentierungsfehler benachteiligt. Deswegen bevorzugen viele Autoren [Mur95, Pös99, Rei04] den *erscheinungsbasierten* Ansatz zur Objekterkennung. Die Merkmalsvektoren werden hier direkt von Pixelwerten im Bild berechnet, wobei keine Segmentierung durchgeführt werden muss. Beide Methoden werden genauer im Kapitel 3 diskutiert.

Wie der Abbildung B.2 entnommen werden kann, manchmal ist es auch für Menschen sehr schwierig, Objekte voneinander zu unterscheiden. Im Allgemeinen können Objekte auf heterogenem Hintergrund plaziert und teilweise verdeckt werden. Unabhängig davon muss ein robustes Objekterkennungssystem in der Lage sein, folgende Probleme handzuhaben:

- *Heterogener Hintergrund*
  In realen Umgebungen befinden sich Objekte oft auf heterogenem Hintergrund (siehe Abbildung B.2). Außerdem können andere Objekte in der direkten Nachbarschaft auftreten,

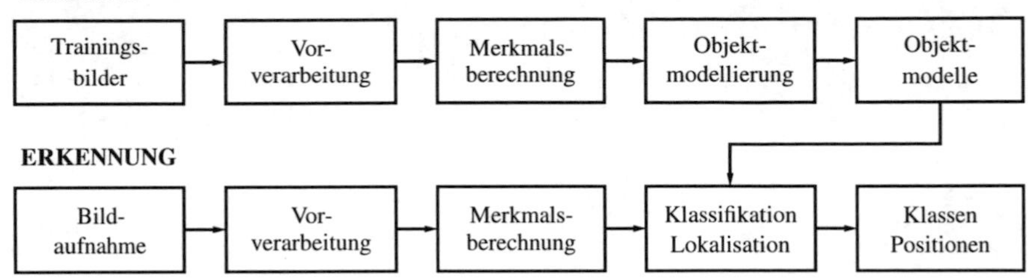

Figure B.3: Ein allgemeines Schema eines Objekterkennungssystems mit der Trainings- und Erkennungsphase [Nie83].

was aufgrund von Schatten und Beleuchtungsänderungen einen negativen Einfluss haben kann. Unabhängig davon muss ein Objekterkennungssystem in der Lage sein, Objekte in einer solchen Umgebung zu finden.

- *Verdeckungen*
  Sehr oft Objekte, die im Bild gesucht werden, sind nicht vollständig sichtbar (siehe Abbildung B.2). Sie sind durch andere Objekte oder Gegenstände verdeckt. In solchen Fällen sollte die Objekterkennung auch fehlerfrei abgewickelt werden.

- *Beleuchtungsänderungen*
  Es kann nicht davon ausgegangen werden, dass die Beleuchtung in realen Umgebungen konstant bleibt. Ein Objekterkennungssystem sollte daher unabhängig von den Beleuchtungsänderungen arbeiten.

- *Mehrobjektszenen*
  Im Allgemeinen ist die Anzahl der Objekte in einem Bild unbekannt. Ein zusätzliches Problem für das Erkennungssystem liegt darin, die Anzahl der Objekte in solchen Szenen zu bestimmen.

In der Abbildung B.3 ist ein allgemeines Schema eines Objekterkennungssystems dargestellt. Ein solches System arbeitet in zwei verschiedenen Betriebsarten, nämlich dem Trainings- und dem Erkennungsmodus. Für das Training werden Bilder aller Objekte aus verschiedenen Ansichten aufgenommen. Dann werden sie vorverarbeitet, und Merkmalsvektoren werden aus ihnen extrahiert. Anschließend werden für alle Objektklassen Ω_κ, die für die Erkennungsaufgabe in Frage kommen, entsprechende Objektmodelle $\mathcal{M}_\kappa$ gelernt. In der Erkennungsphase wird zuerst ein Bild in einer realen Umgebung aufgenommen und vorverarbeitet. Mit dem selben Algorithmus wie in der Trainingsphase werden dann Merkmalsvektoren in dem Bild berechnet. Schließlich startet das System einen Algorithmus, der die extrahierten Merkmalsvektoren mit den gelernten Objektmodellen vergleicht. Als Ergebnis dieses Vergleichs werden die Klassen der Objekte, die im Bild gefunden wurden, und ihre Positionen herausgegeben [Nie83]. Eine genauere Beschreibung all dieser Schritte folgt im weiteren Verlauf der Arbeit.

Die Klassifikation und Lokalisation von Objekten kann in verschiedenen Einsatzgebieten sehr nützlich und manchmal sogar nötig sein. Verfahren zur automatischen computerbasierten Objekterkennung finden ihren Einsatz in u. a. folgenden Gebieten:

- *Gesichtserkennung [Gro04, Ter04]*
  Verschiedene Systeme werden eingesetzt, die Personen anhand von Gesichtern klassifizieren. Solche Systeme sind in der Lage, innerhalb von Sekunden eine Personen in einer großen Datenbank zu finden.

- *Fingerabdruck-Klassifikation [Zha04, Par05]*
  Heutzutage wäre Kriminologie ohne computerbasierte Fingerabdruck-Klassifikation kaum vorstellbar.

- *Handschrifterkennung [Cho04, Heu04]*
  Es gibt viele Situationen, in denen ein automatisches Lesen, Interpretieren und Verarbeiten von handgeschriebenen Texten sehr nützlich ist. Denken Sie an die enorme Anzahl verschiedener Formulare, die jeder von uns in seinem Leben auszufüllen hat!

- *Servicerobotik [You03, Zob03]*
  Serviceroboter müssen in der Lage sein, ein bestimmtes Objekt nicht nur zu finden (klassifizieren), aber auch zu lokalisieren. Um ein Objekt in die Hand zu nehmen, muss man genau wissen, wo sich das Objekt befindet.

- *Medizin [Ben02, Li02]*
  Computer können den Ärzten beim Diagnostizieren sehr behilflich sein. Ein Großteil der medizinischen Daten existiert in Form von digitalen Bildern. Spezielle Erkennungssysteme werden eingesetzt, die in solchen Bildern nach pathologischen Regionen suchen und sowohl dem Arzt als auch dem Patienten die erste Warnung liefern.

- *Visuelle Inspektion [Kum03, Nga05]*
  Oft werden Objekterkennungssysteme eingesetzt, um eine Qualitätsprüfung von seriell produzierten Bauteilen an Produktionsfließbändern durchzuführen.

Obwohl für manche Aufgaben, wie z. B. Fingerabdruck-Klassifikation, robuste Objekterkennungssysteme bereits entwickelt wurden, gibt es immer noch genügend Einsatzgebiete, in denen solche Systeme sehr nützlich wären und unser Leben wesentlich vereinfachen würden.

B.3.2 Beitrag der vorliegenden Arbeit

Diese Arbeit ist eine Fortsetzung der Dissertation von Dr. Michael Reinhold. Dr. Reinhold präsentierte in [Rei04] ein System für erscheinungsbasierte probabilistische Objekterkennung, das Objekte mit lokalen Merkmalen beschreibt. Das erste Ziel der vorliegenden Arbeit liegt in der Erhöhung der Erkennungsraten im Vergleich zu [Rei04] für 3D Szenen mit realem heterogenem Hintergrund. Das in [Rei04] beschriebene System wurde mit künstlich erzeugten Testbildern evaluiert. Die Objekte wurden auf dunklem homogenem Hintergrund aufgenommen und erst später in einen heterogenen Hintergrund kopiert. Auf diese Weise wurde es nicht in Betracht gezogen, dass Objekte abhängig von der Nachbarschaft in dem Aufnahmeprozess ihr Aussehen verändern. In der vorliegenden Arbeit wurden alle Experimente für Testbilder mit realem heterogenem Hintergrund durchgeführt. Außerdem werden in dieser Arbeit Kontextabhängigkeiten zwischen den Objekten in Mehrobjektszenen in Betracht gezogen. In [Rei04] wurden Objekte unabhängig voneinander in Mehrobjektszenen betrachtet. Das Auftreten von bestimmten Objekten wird jedoch in manchen Umgebungen wahrscheinlicher als in anderen.

Die folgenden Hauptaspekte sind neu in der vorliegenden Arbeit:

- *Fusion von mehreren Ansichten*
  Ein neues Verfahren für die Fusion von mehreren Ansichten, das auf der rekursiven Dichtenpropagierung basiert, wird eingeführt. Im Gegensatz zu passiven Ansätzen, in denen die Entscheidung über die Klasse und die Position eines Objektes mit Hilfe eines einzigen Bildes getroffen wird, werden in diesem Algorithmus mehrere Bilder verwendet. Diese Bilder bringen zusätzliche Informationen über die Szene und die beobachteten Objekte mich sich. Die erzielten experimentellen Ergebnisse zeigen, dass, besonders unter schwierigen Bedingungen, die Robustheit des Systems mit dem Fusionsalgorithmus wesentlich steigt.

- *Schnelles Training*
  Ein neues Verfahren wurde entwickelt, bei dem die Bildakquisition für die Trainingsphase mit einer handgeführten Kamera durchgeführt wird. Die fehlenden Objektpositionen werden mit einem Algorithmus berechnet, der in der deutschsprachigen Literatur "Struktur aus Bewegung" [Hei04] genannt wird. Der Lernprozess kann dadurch in einer realen

Umgebung durchgeführt werden. Er ist aber mit einer zusätzlichen Ungenauigkeit der Positionsschätzung verbunden.

- *Kombination von Auflösungsstufen der Wavelet Transformation*
  Die lokalen Merkmalsvektoren werden in der vorliegenden Arbeit auf drei verschiedenen Auflösungsstufen der Wavelet Transformation extrahiert. Dementsprechend können drei Objektmodelle für jedes Objekt in der Trainingsphase erstellt werden. Der Algorithmus zur Klassifikation und Lokalisation von Objekten kann eine Kombination von diesen statistischen Modellen in der Erkennungsphase verwenden. Die Erkennungsraten werden dadurch wesentlich erhöht.

- *Farbmodellierung*
  In [Rei04] werden Objektmodelle ausschließlich mit Hilfe von Grauwertbildern trainiert. Diese Arbeit erweitert die statistische Modellierung, indem auch Farbbilder für das Training verwendet werden.

- *Mehrobjektszenen mit Kontextmodellierung*
  In [Rei04] werden keine Kontextabhängigkeiten zwischen den Objekten in Mehrobjektszenen betrachtet. Die vorliegende Arbeit führt die Kontextmodellierung für Mehrobjektszenen ein.

- *3D-REAL-ENV*
  Im Experimentenkapitel von [Rei04] wurde das System mit Hilfe der DIROKOL Stichprobe [Rei01] mit künstlich erzeugten Testbildern evaluiert. Diese Arbeit präsentiert Ergebnisse für Objekte, die unter realen Bedingungen auf heterogenem Hintergrund aufgenommen wurden. Dazu wurde eine neue Stichprobe erstellt, die mehr als 5000 Testbilder mit realem heterogenem Hintergrund umfasst. Die meisten in dieser Arbeit eingeführten Algorithmen wurden mit dieser 3D-REAL-ENV Stichprobe ausgewertet, wodurch ihr Vergleich besonders objektiv ist.

B.3.3 Motivation

Die im Vergleich zu [Rei04] neuen Komponente des Systems wurden im vorherigen Abschnitt beschrieben. Dieser Abschnitt stellt eine Motivation für diese neuen Komponente vor. Außerdem wird hier die Wahl des statistischen erscheinungsbasierten Ansatzes mit lokalen Merkmalsvektoren begründet.

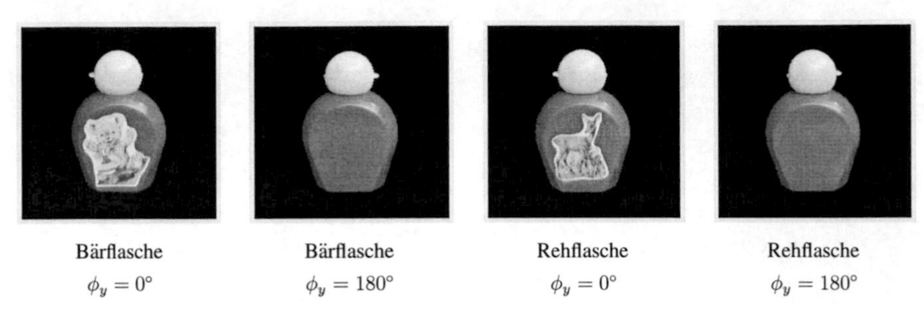

Figure B.4: Zwei verschiedene Objekte, die aus mehreren Ansichten nicht zu unterscheiden sind, z. B. $\phi_y = 180°$.

- *Fusion von mehreren Ansichten*
  Bei manchen Problemen der Objekterkennung können auch Menschen Objekte voneinander nicht unterscheiden. Abbildung B.4 zeigt zwei Flaschen, die aus mehreren Winkeln der externen Rotation ϕ_y identisch aussehen. In solchen Fällen ist es meistens nicht möglich, eine korrekte Klassifikation und Lokalisation anhand von einem einzigen Bild durchzuführen. Aus diesem Grund wird in dieser Arbeit ein Algorithmus zur Fusion von mehreren Ansichten, der auf dem sogenannten Kondensationsalgorithmus [Isa98] basiert, vorgestellt. Die Entscheidung über die Klasse und die Position von einem Objekt wird mit Hilfe mehrerer Bilder getroffen.

- *Schnelles Training*
  Der Lernprozess von Objekterkennungssystemen beginnt mit der Aufnahme von Objekten aus verschiedenen Ansichten. Meistens wird dieser Bildaufnahmeprozess unter Laborbedingungen mit einem Drehteller und einem Kameraarm durchgeführt. In realen Umgebungen ist es aber viel leichter, Objekte mit einer handgeführten Kamera aufzunehmen. Denken Sie an sehr große oder sich ständig bewegende Objekte wie Tiere! Allerdings fehlen bei einer solchen Aufnahmetechnik die Positionsparameter der Objekte in den Bildern. In dieser Arbeit werden diese fehlenden Parameter in einem zusätzlichen Schritt (Struktur aus Bewegung [Hei04]) geschätzt.

- *Kombination von Auflösungsstufen der Wavelet Transformation*
  In [Rei04] werden Objektmerkmale auf einer einzigen Auflösung der Wavelet Transformation berechnet. Die vorliegende Arbeit kombiniert verschiedene Auflösungsstufen der

Wavelet Analyse. Die auf der niedrigen Auflösung erzielten Ergebnisse werden auf einer höheren Auflösung in einem rekursiven Prozess verfeinert. Dieses rekursive Verfahren erhöht die Wahrscheinlichkeit, eine korrekte Klasse und Position zu ermitteln.

- *Farbmodellierung*
  Viele bekannten Systeme zur Klassifikation und Lokalisation verwenden keine Farbinformation von Objekten. Dabei sieht man im alltäglichen Leben, wie oft Menschen die Farbinformation ausnutzen, um Objekte voneinander zu unterscheiden. Manchmal zwei oder mehrere Objekte, die total unterschiedliche Farben haben, sehen sehr ähnlich im Grauwertraum aus. Es ist daher selbstverständlich, dass die Farbe als eine zusätzliche Informationsquelle in ein robustes Objekterkennungssystem hinein gehört.

- *Mehrobjektszenen mit Kontextmodellierung*
  In [Rei04] werden keine Kontextabhängigkeiten zwischen den Objekten in einer Mehrobjektszene in Betracht gezogen. Dabei gibt es viele Situationen, in denen die Kontextinformation über eine bestimmte Szene dem System helfen kann, Objekte zu klassifizieren und zu lokalisieren. Die vorliegende Arbeit modelliert die Kontextzusammenhänge zwischen Objekten in Mehrobjektszenen.

- *3D-REAL-ENV*
  Um eine objektive Auswertung der in der vorliegenden Arbeit präsentierten Algorithmen zu ermöglichen, wurde eine Stichprobe für 3D Objekterkennung in realen Umgebung (3D-REAL-ENV) erstellt. Diese Stichprobe umfasst über 5000 Testbilder mit realem heterogenem Hintergrund.

- *Statistisches Verfahren*
  Das Ziel eines Objekterkennungssystems ist die Klassifikation und Lokalisation von Objekten in einer realen Umgebung. In einer solchen Umgebung hängt das Objektaussehen von vielen Faktoren (Hintergrund, Verdeckungen, Beleuchtungsänderungen) ab. Ein einfacher Vergleich der im Training extrahierten Objektmerkmale mit den Merkmalsvektoren aus einer realen Testszene bringt sehr schlechte Erkennungsresultate. Die einzige Möglichkeit für eine robuste Objekterkennung in realen Umgebungen liegt in der statistischen Modellierung der Merkmalsvektoren in der Trainingsphase. Deshalb werden in dieser Arbeit die Objektmodelle statistisch modelliert, und die Merkmalsvektoren als gaußsche Dichtefunktionen gelernt. Somit können Artefakte und Beleuchtungsänderungen gehandhabt werden.

- *Erscheinungsbasiertes Verfahren*
  Das Rechnersehen unterscheidet zwei Hauptverfahren zur automatischen computerbasierten Objekterkennung. Das Erste basiert auf Ergebnissen eines Segmentierungsschrittes und nutzt die Form des Objektes für seine Beschreibung aus (formbasiert). Das Zweite verwendet die Textur von Objekten bei der Merkmalsberechnung und die Form wird vernachlässigt (erscheinungsbasiert). Die Segmentierung bei dem ersten Ansatz liefert geometrische Merkmale wie Kannten oder Ecken. Diese Merkmale sowie Zusammenhänge zwischen ihnen werden dann bei der Objekterkennung eingesetzt [Hor96]. Allerdings weist Segmentierung zwei wesentliche Nachteile auf; häufige Segmentierungsfehler und Informationsverluste. Im Gegensatz dazu umgehen die erscheinungsbasierten Ansätze diese Probleme. Sie verwenden die Bilddaten, d. h. Pixelwerte, direkt ohne einen Segmentierungsschritt. Zu den bekanntesten erscheinungsbasierten Methoden gehören Schablonen-Vergleich und Eigenraum-Verfahren [Grä03]. Das in der vorliegenden Arbeit beschriebene Objekterkennungssystem soll in der Lage sein, mit verschiedensten Arten von Objekten umzugehen. Es kann nicht davon ausgegangen werden, dass die Form der Objekte bekannt ist. Deswegen wurde das erscheinungsbasierte Verfahren zur Entwicklung des in dieser Arbeit beschriebenen Systems gewählt.

- *Lokale Merkmalsvektoren*
  Es gibt Objekterkennungssysteme, die mit einem einzigen globalen Merkmalsvektor das gesamte Bild beschreiben (z. B. Eigenraum-Verfahren [Grä03]). Es gibt aber auch solche, die mehrere lokale Merkmalsvektoren berechnen, um Objekte zu modellieren (z. B. neuronale Netze [Par04]). Sollte nur ein Pixel im Bild, aufgrund von Artefakten oder Verdeckung, seinen Wert verändern, wird der ganze globale Merkmalsvektor beeinflusst. Wegen dieses Nachteils werden Objekte in dieser Arbeit mit mehreren lokalen Merkmalsvektoren beschrieben.

B.3.4 Überblick

Die Beschreibung mathematischer Aspekte, die bei der Entwicklung des Objekterkennungssystem verwendet wurden, ist im Kapitel 2 zu finden. Das Wissen über die statistischen Grundlagen, die Wavelet Transformation und die Funktionsapproximation ist erforderlich, um die statistische Objektmodellierung verstehen zu können.

Ein Überblick über die existierenden Algorithmen und Methoden zur Objekterkennung wird im Kapitel 3 präsentiert. Angefangen mit einigen grundlegenden Begriffen auf diesem Gebiet beschreibt dieses Kapitel bekannte formbasierte und erscheinungsbasierte Ansätze zur Klassifikation und Lokalisation von Objekten

Die Trainingsphase des Systems wird im Kapitel 4 diskutiert. Zuerst wird die Akquisition von Trainingsdaten beschrieben. Später folgt die Beschreibung der statistischen Modellierung mit den Grauwert- und Farbbildern. Die Kontextmodellierung bei Mehrobjektszenen wird am Ende dieses Kapitels erklärt.

Die Erkennungsphase des Systems wird im Kapitel 5 beschrieben. Zuerst wird das Problem der Klassifikation und Lokalisation von Objekten in Einobjektszenen angesprochen, wonach die Beschreibung des Erkennungsproblems in Mehrobjektszenen folgt.

Durchgeführte Experimente und erzielte Ergebnisse werden im Kapitel 6 vorgestellt. Zuerst wird die experimentelle Umgebung, die Evaluationskriterien und die neue 3D-REAL-ENV Stichprobe beschrieben. Später werden alle im Kapitel 5 eingeführten Algorithmen experimentell ausgewertet und verglichen.

Kapitel 6 schließt die Arbeit mit einer Schlussfolgerung, die aus einer Zusammenfassung und einem Ausblick besteht.

B.4 Zusammenfassung

Eines der grundlegendsten Probleme des Rechnersehens liegt in der Erkennung von Objekten in digitalen Bildern. Dabei verstecken sich hinter dem Begriff Objekterkennung sowohl die Klassifikation als auch die Lokalisation von Objekten. Die Aufgabe der Objektklassifikation besteht darin, die Klassen von Objekten, die in einem Bild $\boldsymbol{f}$ auftreten, aus einer Menge der vordefinierten Objektklassen $\Omega = \{\Omega_1, \Omega_2, \ldots, \Omega_\kappa, \ldots, \Omega_{N_\Omega}\}$ zu bestimmen. Im Allgemeinen ist die Anzahl der Objekte in einer Szene unbekannt. Deshalb ist es erforderlich, die Objektanzahl zu schätzen. Bei der Objektlokalisation ermittelt das System die Positionen von Objekten im Bild, wobei die Objektklassen sind a-priori bekannt. Die Positionen von Objekten sind relativ zueinander mit einem dreidimensionalen Translationsvektor $\boldsymbol{t} = (t_x, t_y, t_z)^{\mathrm{T}}$ und einem dreidimensionalen Rotationsvektor $\boldsymbol{\phi} = (\phi_x, \phi_y, \phi_z)^{\mathrm{T}}$ definiert, wobei der Ursprung des Koordinatensystems für diese Transformationsparameter im Mittelpunkt des Bildes liegt.

Die Verfahren zur Objekterkennung lassen sich in zwei Kategorien aufteilen, nämlich die formbasierten und die erscheinungsbasierten Methoden. Die formbasierten Ansätze führen zuerst einen Segmentierungsschritt durch, um geometrische Merkmale wie Linien oder Ecken für die Repräsentation von Objekten zu extrahieren [Hor96, Lat00, Lat02, Ker03, Che04]. Die bekanntesten formbasierten Objekterkennungssysteme bestehen üblicherweise aus drei Modulen. Das erste Modul ist für die Repräsentation der Form zuständig, das zweite für die Modellierung und das dritte für die Klassifikation und Lokalisation [Win94]. Leider sind diese Methoden oft von Segmentierungsfehlern benachteiligt. Aus diesem Grund bevorzugen viele Autoren [Mur95, Pös99, Rei04] einen anderen, texturbasierten Weg. Die Objektmerkmale werden hier direkt von den Pixelintensitäten berechnet, und eine vorherige Segmentierung ist nicht erforderlich. Zu den bekanntesten Algorithmen zur texturbasierten (erscheinungsbasierten) Klassifikation und Lokalisation von Objekten gehören: Schablonen-Vergleich [Bru97, Gon01, Pra01], Eigenraum-Verfahren [Tur91, Leo96, Mog97, Grä03] und Support Vector Machines [Cor95, Vap95].

Diese Arbeit präsentiert einen erscheinungsbasierten Algorithmus zur Objekterkennung, in dem die Merkmalsvektoren direkt aus Pixelintensitäten berechnet werden. Bei der Modellierung mit Hilfe von Grauwertbildern werden Objekte mit zweidimensionalen lokalen Merkmalsvektoren beschrieben, während bei der Farbmodellierung sechsdimensionale Merkmale extrahiert werden. Der Hauptvorteil der lokalen Merkmalsvektoren liegt darin, dass lokale Artefakte nur einige von ihnen beeinträchtigen, während die restlichen unverändert bleiben. Im Gegensatz dazu kann sich ein globaler Merkmalsvektor total verändern, wenn nur ein einziger Pixelwert variiert. Das System ermittelt eine Menge der lokalen Merkmalsvektoren für alle Trainingsbilder mit Hilfe der Wavelet Transformation. Johnston Wavelet und die entsprechende Skalierungsfunktion kommen dabei zum Einsatz. Im Falle der Grauwertmodellierung resultiert die erste Komponente des Merkmalsvektors aus einer Tiefpassfilterung, während in der zweiten Komponente Informationen über Unstetigkeiten in der kleinen Umgebung gespeichert werden (Hochpassfilterung). Bei der Farbmodellierung ist die Bedeutung der Komponente analog, wobei das erste und zweite Element im roten Kanal, das dritte und vierte Element im grünen Kanal und das fünfte und sechste Element im blauen Kanal berechnet werden.

Manche Merkmalsvektoren $\boldsymbol{c}_{\kappa,m}$ beschreiben das Objekt Ω_κ, andere gehören zum Hintergrund. In einer realen Umgebung kann nicht davon ausgegangen werden, dass der Hintergrund im Voraus bekannt ist. Logischerweise sollten bei der Objektmodellierung nur die Merkmale in Betracht gezogen werden, die tatsächlich das Objekt beschreiben. Da das Objekt üblicherweise nur einen Teil des Bildes annimmt, wird in der Trainingsphase ein sogenanntes Objektfenster O_κ trainiert, das das Objekt Ω_κ sehr eng umfasst. Im Falle von 3D Objekten verändert sich die Position, Orientierung und Größe des Objektfensters vom Bild zum Bild. Deswegen wird es als eine kontinuierliche Funktion der Positionsparameter modelliert $O_\kappa(\boldsymbol{\phi}, \boldsymbol{t})$. Anders ausgedrückt, dank des Objektfensters O_κ ist das System in der Lage, die Menge der Objektmerkmalsvektoren $\boldsymbol{c}_{\kappa,m}$ im gesamten Positionsparameterraum $(\boldsymbol{\phi}, \boldsymbol{t})$ zu ermitteln.

Um Beleuchtungsänderungen und niederfrequente Artefakte handhaben zu können, werden die lokalen Merkmalsvektoren $\boldsymbol{c}_{\kappa,m}$ als Zufallsvektoren interpretiert, wobei die statistische Unabhängigkeit ihrer Komponente angenommen wird. Da man zusätzlich von der statistischen Unabhängigkeit der Objektmerkmalsvektoren von den Hintergrundmerkmalen ausgeht, können Objekte separat von dem Hintergrund modelliert werden. Für alle Merkmalsvektoren $\boldsymbol{c}_{\kappa,m}$, die zum Objektfenster gehören O_κ, werden entsprechende Mittelwertvektoren $\boldsymbol{\mu}_{\kappa,m}$ und Standardabweichungsvektoren $\boldsymbol{\sigma}_{\kappa,m}$ als kontinuierliche Funktionen der Positionsparameter $(\boldsymbol{\phi}, \boldsymbol{t})$ gelernt. Somit werden Objektmerkmalsvektoren mit gaußschen Dichtefunktionen $p(\boldsymbol{c}_m|\boldsymbol{\mu}_{\kappa,m}, \boldsymbol{\sigma}_{\kappa,m}, \boldsymbol{\phi}, \boldsymbol{t})$ repräsentiert. Die Hintergrundmerkmale $\boldsymbol{c}_{\kappa,m}$, die außerhalb des Objektfensters O_κ liegen, werden auch statistisch modelliert, wobei in dem Falle die Gleichverteilung zum Einsatz kommt. Sie werden also als konstante Dichtefunktionen $p(\boldsymbol{c}_m) = p_b$ in den statistischen Objektmodellen gespeichert.

Zusammenfassend, für alle Objektklassen Ω_κ, die in einem bestimmten Objekterkennungsproblem betrachtet werden, werden in der Trainingsphase statistische Objektmodelle $\mathcal{M}_\kappa$ gelernt. Diese Modelle sind als kontinuierliche Funktionen der Positionsparameter $\mathcal{M}_\kappa(\boldsymbol{\phi}, \boldsymbol{t})$ definiert. Das bedeutet, dass jedes Objektmodell $\mathcal{M}_\kappa$ das Objektfenster O_κ, die Menge von Objektmerkmalsvektoren $\boldsymbol{c}_{\kappa,m}$, ihre Dichtefunktionen $p(\boldsymbol{c}_m|\boldsymbol{\mu}_{\kappa,m}, \boldsymbol{\sigma}_{\kappa,m})$, und den Wert der Hintergrunddichte $p(\boldsymbol{c}_m) = p_b$ für alle möglichen Positionsparameter $(\boldsymbol{\phi}, \boldsymbol{t})$ im kontinuierlichen Sinne enthält.

Es ist auch möglich, Kontextzusammenhänge zwischen den Objekten in der Trainingsphase zu lernen. In bekannten Umgebungen ist das Auftreten mancher Objekte wahrscheinlicher als anderer. In diesem Fall trainiert das System die a-priori Wahrscheinlichkeiten $p_\iota(\Omega_\kappa)$ für das Auftreten aller Objekte Ω_κ in vordefinierten Kontexten Υ_ι. Diese diskreten Dichtefunktionen werden in den sogenannten statistischen Kontexmodellen $\mathcal{M}_\iota$ gespeichert.

Nachdem für alle möglichen Objektklassen Ω_κ und alle möglichen Kontexte Υ_ι die entsprechenden Modelle ($\mathcal{M}_\kappa$, $\mathcal{M}_\iota$) bereits in der Trainingsphase gelernt wurden, ist das System in der Lage, Objekte in realen Szenen zu erkennen (klassifizieren und lokalisieren). Dieses reale Testbild wird zuerst vorverarbeitet, und Merkmalsvektoren werden in ihm berechnet. Dann wird einer der vier Erkennungsalgorithmen, die in das System integriert wurden, gestartet. Der erste Algorithmus löst das Problem der Objekterkennung in Einobjektszenen anhand von nur einem Bild. Das zweite Verfahren beschäftigt sich auch mit Einobjektszenen, aber diesmal mehrere Ansichten des Objekts zum Einsatz kommen. Der dritte Ansatz wird für Mehrobjektszenen ohne Kontextabhängigkeiten eingesetzt, während beim vierten Algorithmus die Kontextzusammenhänge zwischen den Objekten in Mehrobjektszenen betrachtet werden.

Die Klassifikation und Lokalisation von Objekten in Einobjektszenen wird mit dem sogenannten ML-Algorithmus (Maximum Likelihood Estimation) gelöst. Zuerst werden lokale Merkmalsvektoren in den vorverarbeiteten Grauwert- oder Farbbildern mit der Wavelet Transformation extrahiert. Dann wird das Objektfenster O_κ für alle Objektklassen- κ und Positionshypothesen $(\boldsymbol{\phi}_h, \boldsymbol{t}_h)$ mit Hilfe der gelernten Objektmodelle $\mathcal{M}_\kappa$ bestimmt. Nur Merkmale $\boldsymbol{c}_m$ innerhalb des Objektfensters werden bei der Auswertung aller Hypothesen in Betracht gezogen. Die Dichtenwerte $p(\boldsymbol{c}_m|\boldsymbol{\mu}_{\kappa,m}, \boldsymbol{\sigma}_{\kappa,m})$ der Objektmerkmalsvektoren $\boldsymbol{c}_m$, die größer als der Wert der Hintergrunddichte p_b sind, werden dann miteinander multipliziert. Das Ergebnis dieses Produkts wird später mit dem sogenannten geometrischen Kriterium normalisiert und anschließend über alle Klassen- und Positionshypothesen maximiert. Auf diese Weise findet das System den Index der optimalen Objektklasse $\widehat{\kappa}$ sowie die optimale Positionshypothese $(\widehat{\boldsymbol{\phi}}, \widehat{\boldsymbol{t}})$ in einem Bild.

Da manche Objekte aus einem einzigen Winkel voneinander nicht unterschieden werden können, verwendet der zweite in das System integrierte Algorithmus zur Objekterkennung eine Fusion von mehreren Objektansichten $\langle \boldsymbol{f} \rangle_n = (\boldsymbol{f}_0, \boldsymbol{f}_1, \ldots, \boldsymbol{f}_n)$, wobei als Ergebnis der Klassifikation immer noch exakt eine Objektklasse $\Omega_{\widehat{\kappa}}$ erwartet wird. Die Klasse zusammen mit der Position eines Objektes im Bild $\boldsymbol{f}_n$ werden in diesem Verfahren als Objektzustand $\boldsymbol{q}_n$ bezeichnet. Die Fusion von mehreren Ansichten definiert das Objekterkennungsproblem mit der Schätzung des Objektzustands $\boldsymbol{q}_n$ für die gegebene Bildsequenz $\langle \boldsymbol{f} \rangle_n$ und Kamerabewegungen $\langle \boldsymbol{a} \rangle_{n-1}$ zwischen ihnen. Die Aufgabe wird mit der Maximierung der Objektzustandsdichte $p(\boldsymbol{q}_n|\boldsymbol{f}_n, \boldsymbol{a}_{n-1}, \ldots, \boldsymbol{a}_0, \boldsymbol{f}_0)$, die rekursiv mit dem Kondensationsalgorithmus propagiert wird, gelöst.

Der dritte Ansatz adressiert das Problem der Klassifikation und Lokalisation von Objekten in Mehrobjektszenen, wobei keine Kontextzusammenhänge betrachtet werden. In diesem Fall müssen nicht nur die Objektklassen und Positionen bestimmt werden. Da die Anzahl der Objekte $\widehat{\imath}$ in einer Mehrobjektszene unbekannt ist, muss diese zuerst ermittelt werden. Bei dem Algorithmus nimmt das System an, dass die a-priori Wahrscheinlichkeit für das Auftreten aller Objekte Ω_κ gleich ist. Deshalb kann dazu die ML-Schätzung (Maximum Likelihood Estimation) eingesetzt werden. Die ersten $\widehat{\imath}$ meist wahrscheinlichen Objektklassen und ihre Positionen werden dann als das endgültige Erkennungsergebnis betrachtet.

Das letzte in das System integrierte Erkennungsverfahren beschäftigt sich auch mit dem Objekterkennungsproblem in Mehrobjektszenen, wobei diesmal die Kontextabhängigkeiten zwischen den Objekten in Betracht gezogen werden. Die Kontextzusammenhänge wurden in der Trainingsphase statistisch modelliert. Da es kein a-priori Wissen über den Kontext, in dem eine Szene aufgenommen wurde, gibt, arbeitet der Algorithmus in zwei Schritten. Zuerst wird der Kontext $\Upsilon_{\widehat{\imath}}$ des Testbildes automatisch bestimmt. Dann erkennt das System Objekte in dem Bild, wobei die statistischen Kontextmodelle $\mathcal{M}_{\hat{\iota}}$, die in der Trainingsphase gelernt wurden, hier zum Einsatz kommen.

Um die Experimente objektiv durchzuführen, wurde eine Stichprobe von Bildern für 3D Objekterkennung in realen Umgebungen (3D-REAL-ENV) erstellt. Die Stichprobe besteht aus zehn Objekten. Die Trainingsbilder von den Objekten wurden auf dunklem homogenem Hintergrund unter zwei unterschiedlichen Beleuchtungen aus 1680 Ansichten aufgenommen. Deshalb gibt es insgesamt 33600 Trainingsszenen. Für die Testphase wurden drei Arten von Testbildern aus 288 verschiedenen Ansichten aufgenommen. Es waren 2880 Bilder mit homogenem Hintergrund, 2880 Bilder mit leicht heterogenem Hintergrund, und 2880 Bilder mit sehr heterogenem Hintergrund. Die Beleuchtung bei den Testbildern ist unterschiedlich von der Beleuchtung bei den Trainingsbilder. Die Testansichten sind im Allgemeinen auch unterschiedlich von den Trainingsansichten. Zusätzlich wurden mehr als 200 verschiedene reale Hintergründe bei den Aufnahmen verwendet. Aufgrund von diesen Eigenschaften ist das Problem der Klassifikation und Lokalisation von Objekten mit der 3D-REAL-ENV Stichprobe sehr schwierig.

Die Klassifikations- und Lokalisationsraten sowie andere experimentelle Ergebnisse, die mit dem in der vorliegenden Arbeit beschriebenen System erzielt wurden, beweisen eine hohe Robustheit des Objekterkennungssystems in realen Umgebungen. Die zwei Hauptbeiträge dieser Arbeit, nämlich die Farb- und Kontextmodellierung, erhöhen diese Robustheit wesentlich. Bei den 3D-REAL-ENV Testbildern mit sehr heterogenem Hintergrund beträgt die Klassifikationsrate für die Grauwertmodellierung 54.1% und konnte mit der Farbmodellierung auf 82.3% erhöht werden. Die Lokalisation wurde mit dem Einsatz der Farbinformation auch verbessert, und zwar von 69.0% (Grauwertmodellierung) auf 73.6% (Farbmodellierung). Außerdem, dank der Kontextmodellierung, wurde die Klassifikation in Mehrobjektszenen wesentlich robuster. Die Klassifikationsrate in Mehrobjektszenen mit sehr heterogenem Hintergrund beträgt 62.9% ohne Kontextmodellierung. Nach dem Einsatz der Kontextmodellierung wurde sie auf 87.5% erhöht.

Appendix C

Mathematical Symbols

Ω	Set of all object classes	1
κ	Index for object classes	1
Ω_κ	Object class number κ	1
N_Ω	Number of all object classes	1
$\boldsymbol{t} = (t_x, t_y, t_z)^{\mathrm{T}}$	Translation vector	2
$\boldsymbol{\phi} = (\phi_x, \phi_y, \phi_z)^{\mathrm{T}}$	Rotation vector	2
$\boldsymbol{t}_{\text{int}} = (t_x, t_y)^{\mathrm{T}}$	Internal translation vector	2
$\phi_{\text{int}} = \phi_z$	Internal rotation	2
$t_{\text{ext}} = t_z$	External translation	2
$\boldsymbol{\phi}_{\text{ext}} = (\phi_x, \phi_y)^{\mathrm{T}}$	External rotation vector	2
$\mathcal{M}$	Set of all object models	4
$\mathcal{M}_\kappa$	Object model number κ	4
p	Density	13
δ	Impulse function	13
$\boldsymbol{\Sigma}$	Covariance matrix	15
j	Imaginary unit	19
FT	Fourier transform	20
L^2	Hilbert space	21

ψ	Wavelet – basis function for wavelet transformation	21
Ψ	Fourier transform for wavelet	21
WT	Wavelet transform	22
s	Multiresolution scale parameter for wavelet transformation	22
k	Multiresolution shift parameter for wavelet transformation	22
φ	Scaling function for wavelet transformation	23
$\boldsymbol{c}$	Feature vector	41
$N_{\boldsymbol{c}}$	Number of feature vector components	41
$\boldsymbol{f}$	Image	43
C	Set of feature vectors	43
N_C	Number of feature vectors	43
m	Index for feature vectors	43
ρ	Index for training images	44
N_ρ	Number of training images	44
d	Distance function	45
$\widehat{\kappa}$	Index of resulting object class after classification	46
$\widehat{\boldsymbol{\phi}}$	Resulting rotation vector after localization	46
$\widehat{\boldsymbol{t}}$	Resulting translation vector after localization	46
O	Object area	53
$\boldsymbol{x}$	Grid point for feature extraction	53
$\widehat{\phi}_{\text{int}}$	Resulting internal rotation after localization	56
$\widehat{\boldsymbol{\phi}}_{\text{ext}}$	Resulting external rotation vector after localization	56
$\widehat{\boldsymbol{t}}_{\text{int}}$	Resulting internal translation vector after localization	56
$\widehat{t}_{\text{ext}}$	Resulting external translation after localization	56
cov	Covariance	58
$\boldsymbol{u}$	Eigen vector	58

λ	Eigen value	58
$\boldsymbol{\Gamma}$	Measurement matrix for structure-from-motion algorithm	68
N_F	Number of frames	68
$\boldsymbol{S}$	Structure matrix for structure-from-motion algorithm	69
$\boldsymbol{M}$	Motion matrix for structure-from-motion algorithm	69
$\boldsymbol{P}$	Camera projection matrix	70
$\boldsymbol{K}$	Matrix for camera intrinsic parameters	70
$\boldsymbol{R}$	Camera rotation matrix	70
$\boldsymbol{t}_c$	Camera translation vector	70
$\widehat{s}$	Minimum scale for wavelet transformation	72
X	Set of grid points for feature extraction	73
$\widehat{M}$	Number of grid points for feature extraction	73
ξ	Assignment function for feature vectors	78
$\boldsymbol{\mu}$	Mean vector	82
$\boldsymbol{\sigma}$	Standard deviation vector	82
$\boldsymbol{B}$	Matrix for statistical parameters	82
p_b	Background density value	86
Υ	Set of contexts	93
ι	Index for contexts	93
N_Υ	Number of contexts	93
N_ι	Number of images for a context	93
η	Normalization factor	93
$\mathcal{M}_\iota$	Context model number ι	95
Q	Quality measure	105
$\boldsymbol{a}$	Camera movement for fusion of multiple views	109
$\boldsymbol{q}$	Object state for fusion of multiple views	109

N_h	Number of hypotheses for maximization	111
$\widehat{\boldsymbol{q}}$	Resulting object state after fusion of multiple views	111
Z	Set of particles for condensation algorithm	112
c_r	Classification rate	123
l_r	Localization rate	123
d_r	Determination rate	124
I_{lum}	Type of illumination	125
T_{ype}	Type of test images	127
N_P	Number of object parts	153

List of Figures

List of Tables

Bibliography

[Alo90] J. Y. Aloimonos. Perspective approximations. *Image and Vision Computing*, 8(3):177–192, August 1990.

[Bam72] G. Bamberg. *Statistische Entscheidungstheorie*. Physica-Verlag, Würzburg, Germany, 1972.

[Bar81] A. H. Barr. Superquadrics and angle-preserving transformations. *IEEE Computer Graphics and Applications*, 1(1):11–23, January 1981.

[Ben02] Y. Bentoutou, N. Taleb, M. Chikr El Mezouar, M. Taleb, and L. Jetto. An invariant approach for image registration in digital subtraction angiography. *Pattern Recognition*, 35(12):2853–2865, December 2002.

[Bla06] I. Blayvas, A. Bruckstein, and R. Kimmel. Efficient computation of adaptive threshold surfaces for image binarization. *Pattern Recognition*, 39(1):89–101, Janurary 2006.

[Bri70] C. R. Brice and C. L. Fenema. Scene analysis using regions. *Artificial Intelligence*, 1:205–226, 1970.

[Bro85] I.N. Bronstein and K. A. Semendjajew. *Taschenbuch der Mathematik*. Harri Deutsch, Thun, Germany, 1985.

[Bru97] R. Brunelli and T. Poggio. Template matching: Matched spatial filters and beyond. *Pattern Recognition*, 30(5):751–768, May 1997.

[BS88] Y. Bar-Shalom and T. E. Fortmann. *Tracking and Data Association*. Academic Press, Boston, San Diego, New York, 1988.

[Can86] J. Canny. A computational approach to edge detection. *IEEE Trans. Pattern Analysis and Machine Intelligence*, PAMI-8(6):679–698, November 1986.

[Cas96] K. R. Castleman. *Digital Image Processing*. Prentice Hall, New Jersey, USA, 1996.

[Ces05] R. Cesar, E. Bengoetxea, I. Bloch, and P. Larranaga. Inexact graph matching for model-based recognition: Evaluation and comparison of optimization algorithms. *Pattern Recognition*, 38(11):2099–2113, November 2005.

[Che66] W. Cheney. *Introduction to Approximation Theory*. McGraw-Hill Book Company, New York, USA, 1966.

[Che00] W. Cheney and W. Light. *A Course in Approximation Theory*. Brooks/Cole Publishing, Pacific Grove, USA, 2000.

[Che04] H. Chen, I. Shimshoni, and P. Meer. Model based object recognition by robust information fusion. In *17th International Conference on Pattern Recognition*, Cambrige, UK, August 2004.

[Cho04] Sung-Jung Cho and Jin H. Kim. Bayesian network modeling of strokes and their relationships for on-line handwriting recognition. *Pattern Recognition*, 37(2):253–264, February 2004.

[Chr00] N. Christianini and J. Shawe-Taylor. *An Introduction to Support Vector Machines*. Cambridge University Press, Cambridge, UK, 2000.

[Chu92] C. Chui. *An Introduction to Wavelets*. Academic Press, San Diego, USA, 1992.

[Cor95] C. Cortes and V. N. Vapnik. Support vector networks. *Machine Learning*, 20:273–297, 1995.

[Dah99] J. Dahmen, R. Schlüter, and H. Ney. Discriminative training of gaussian mixtures for image object recognition. In W. Förstner, J. Buhmann, A. Faber, and P. Faber, editors, *21. DAGM Symposium Mustererkennung*, pages 205–212, Bonn, Germany, September 1999. Springer-Verlag.

[Dau90] I. Daubechies. The wavelet transform, time-frequency localization, and signal analysis. *IEEE Transactions on Information Theory*, 36(9):961–1005, September 1990.

[Dav75] P. J. Davis. *Interpolation and Approximation*. Dover Publications, New York, USA, 1975.

[Dei05] F. Deinzer. *Optimale Ansichtenauswahl in der aktiven Objekterkennung*. Logos Verlag, Berlin, Germany, 2005.

[dV98] V. C. de Verdiere and J. L. Crowley. Visual recognition using local appearance. In H. Burkhardt and B. Neumann, editors, *Proceedings of the 5th European Conference on Computer Vision*, pages 640–654, Freiburg, Germany, Juni 1998. Springer-Verlag.

[EP02] M. Egmont-Petersen, D. de Ridder, and H. Handels. Image processing with neural networks – a review. *Pattern Recognition*, 35(10):2279–2301, October 2002.

[Fau92] O. D. Faugeras, Q.-T. Luong, and S. J. Maybank. Camera self-calibration: Theory and experiments. In G. Sandini, editor, *Proceedings of the 2nd European Conference on Computer Vision*, pages 321–334, Santa Margherita Ligure, Italy, May 1992. Springer-Verlag.

[För87] W. Förstner and E. Gülch. A fast operator for detection and precise location of distinct points, corners, and centers of circular features. In *Intercommission Conference on Fast Processing of Photogrammetric Data*, pages 281–305, Interlaken, Switzerland, September 1987.

[Frü04] J. Fründ, J. Gausemeier, C. Matysczok, and R. Radkowski. Using augmented reality technology to support the automobile development. In W. Shen, Z. Lin, J.-P. A. Barthès, and T. Li, editors, *8th International Conference on Computer Supported Cooperative Work in Design*, pages 289–298, Xiamen, China, May 2004. Springer-Verlag.

[Fu81] K. S. Fu and J. K. Mui. A survey on image segmentation. *Pattern Recognition*, 13(1):3–16, January 1981.

[Fus00] A. Fusiello. Uncalibrated euclidean reconstruction: a review. *Image and Vision Computing*, 18(6-7):555–563, May 2000.

[Fus01] A Fusiello. A new autocalibration algorithm: Experimental evaluation. In W. Skarbek, editor, *9th International Conference on Computer Analysis of Images and Patterns*, pages 717–724, Warsaw, Poland, September 2001. Springer-Verlag.

[Gau05] J. Gausemeier, M. Grafe, C. Matysczok, R. Radkowski, J. Krebs, and H. Oelschlaeger. Eine mobile augmented reality versuchsplattform zur untersuchung und evaluation von fahrzeugergonomien. In T. Schulze, G. Horton, B. Preim, and S. Schlechtweg, editors, *Simulation und Visualisierung*, pages 185–194, Magdeburg, Germany, March 2005. SCS Publishing House e.V.

[God89] G. D. Godin and M. D. Levine. Structured edge map of curved objects in a range image. In *Proceedings of the Conference on Computer Vision and Pattern Recognition*, pages 276–281, San Diego, USA, June 1989.

[Gon92] R. C. Gonzales and R. W. Woods. *Digital Image Processing*. Addison-Wesley Longman Publishing Co., Boston, USA, 1992.

[Gon01] R. C. Gonzalez and R. E. Woods. *Digital Image Processing*. Prentice Hall, New Jersey, USA, 2001.

[Gor90] K. E. Gorlon, S. Orlow, and P. S. Plexico. *Data Abstraction and Object-Oriented Programming in C++*. John Wiley & Sons, Chichester, USA, 1990.

[Grä03] Ch. Gräßl, F. Deinzer, and H. Nieman. Continuous parametrization of normal distribution for improving the discrete statistical eigenspace approach for object recognition. In V. Krasnoproshin, S. Ablameyko, and J. Soldek, editors, *Pattern Recognition and Information Processing 03*, pages 73–77, Minsk, Belarus, Mai 2003.

[Gro04] R. Gross, I. Matthews, and S. Baker. Appearance-based face recognition and light-fields. *IEEE Transactions on Pattern Analysis and Machine Intelligence*, 26(4):449–465, April 2004.

[Grz03] M. Grzegorzek, F. Deinzer, M. Reinhold, J. Denzler, and H. Niemann. How fusion of multiple views can improve object recognition in real-world environments. In T. Ertl, B. Girod, G. Greiner, H. Niemann, H.-P. Seidel, E. Steinbach, and R. Westermann, editors, *Vision, Modeling, and Visualization 2003*, pages 553–560, Munich, Germany, November 2003. Aka/IOS Press, Berlin, Amsterdam.

[Grz04a] M. Grzegorzek, K. Pasumarthy, M. Reinhold, and H. Niemann. Statistical object recognition for multi-object scenes with heterogeneous background. In B. Chanda, S. Chandran, and L. Davis, editors, *4th Indian Conference on Computer Vision, Graphics and Image Processing*, pages 222–227, Kolkata, India, December 2004. Allied Publishers Private Limited, Kolkata.

[Grz04b] M. Grzegorzek, I. Scholz, M. Reinhold, and H. Niemann. Fast training for object recognition with structure-from-motion. In V.V. Geppener, I.B. Gurevich, S.E.

Ivanova, A.P. Nemirko, H. Niemann, D.V. Puzankov, Yu.O. Trusova, and Yu.I. Zhuravlev, editors, *7th International Conference on Pattern Recognition and Image Analysis 2004: New Information Technologies*, pages 231–234, St. Petersburg, Russia, October 2004. SPbETU, St. Petersburg.

[Grz05a] M. Grzegorzek and H. Niemann. Statistical object recognition including color modeling. In M. Kamel and A. Campilho, editors, *2nd International Conference on Image Analysis and Recognition*, pages 481–489, Toronto, Canada, September 2005. Springer-Verlag, Berlin, Heidelberg, LNCS 3656.

[Grz05b] M. Grzegorzek, M. Reinhold, and H. Niemann. Feature extraction with wavelet transformation for statistical object recognition. In M. Kurzynski, E. Puchala, M. Wozniak, and A. Zolnierek, editors, *4th International Conference on Computer Recognition Systems*, pages 161–168, Rydzyna, Poland, May 2005. Springer-Verlag, Berlin, Heidelberg.

[Grz05c] M. Grzegorzek, I. Scholz, M. Reinhold, and H. Niemann. Fast training for object recognition with structure-from-motion. *Pattern Recognition and Image Analysis: Advances in Mathematical Theory and Applications*, 15(1):183–186, January 2005.

[Har69] R. M. Haralick and G. L. Kelly. Pattern recognition with measurement space and spatial clustering for multiple images. *Proc. IEEE*, 57(4):654–665, April 1969.

[Har83] R. M. Haralick. Ridges and valleys on digital images. *Computer Vision, Graphics, and Image Processing*, 22(10):28–38, April 1983.

[Har85] R. M. Haralick and L. G. Shapiro. Image segmentation techniques. *Computer Vision, Graphics, and Image Processing*, 29(1):100–132, January 1985.

[Har88] C. Harris and M. Stephens. A combined edge and corner detector. In *4th Alvey Vision Conference*, pages 189–192, Manchester, UK, August 1988.

[Har93] R. Hartley. Euclidean reconstruction from uncalibrated views. In *Applications of Invariance in Computer Vision*, pages 237–256. Springer-Verlag, 1993.

[Har03] R. Hartley and A. Zisserman. *Multiple View Geometry in Computer Vision*. Cambridge University Press, Cambridge, UK, 2003.

[Hei04] B. Heigl. *Plenoptic Scene Modeling from Uncalibrated Image Sequences*. ibidem-Verlag, Stuttgart, Germany, 2004.

[Heu04] L. Heutte, A. Nosary, and T. Paquet. A multiple agent architecture for handwritten text recognition. *Pattern Recognition*, 37(4):665–674, April 2004.

[Hor86] B. Horn. *Robot Vision*. MIT Press and McGraw-Hill, Cambridge, USA, 1986.

[Hor96] J. Hornegger. *Statistische Modellierung, Klassifikation und Lokalisation von Objekten*. Shaker Verlag, Aachen, Germany, 1996.

[Hua06] L.-L. Huang and A. Shimizu. A multi-expert approach for robust face detection. *Pattern Recognition*, 39:To Appear, 2006.

[Isa98] M. Isard and A. Blake. Condensation – conditional density propagation for visual tracking. *International Journal of Computer Vision*, 29(1):5–28, January 1998.

[Jac80] C. L. Jackins and S. L. Tanimoto. Octtrees and their use in representing three-dimensional objects. *Computer Graphics and Image Processing*, 14(3):249–270, November 1980.

[Kal60] R.E. Kalman. A new approach to linear filtering and prediction problems. *Journal of Basic Engineering*, pages 35–44, 1960.

[Kan98] T. Kanade and D. D. Morris. Factorization methods for structure from motion. *Philosophical Transactions: Mathematical, Physical and Engineering Sciences*, 356(1740):1153–1173, May 1998.

[Kar46] K. Karhunen. Zur spektraltheorie stochastischer prozesse. *Annales Academiae Scientiarum Fennicae*, 37:1–37, 1946.

[Ker03] J. Kerr and P. Compton. Toward generic model-based object recognition by knowledge acquisition and machine learning. In *Proceedings of the Eighteenth International Joint Conference on Artificial Intelligence*, pages 9–15, Acapulco, Mexico, August 2003.

[Koe79] J. J. Koenderik and A. J. Van Doorn. The internal representation of solid shape with respect to vision. *Biological Cybernetics*, 32:211–216, 1979.

[Kra95] P. Kral. Wavelet transforms. Technical report, Informatik 5 (Lehrstuhl für Mustererkennung), Universität Erlangen-Nürnberg, 1995.

[Kum03] A. Kumar. Neural network based detection of local textile defects. *Pattern Recognition*, 36(7):1631–1644, July 2003.

[Lat00] L. J. Latecki, R. Lakaemper, and D. Wolter. Shape similarity measure based on correspondence of visual parts. *Pattern Analysis and Machine Intelligence*, 22(10):1185–1190, October 2000.

[Lat02] L. J. Latecki and R. Lakaemper. Application of planar shape comparison to object retrieval in image databases. *Pattern Recognition*, 35(1):15–19, January 2002.

[Lat05] L. J. Latecki, R. Lakaemper, and D. Wolter. Optimal partial shape similarity. *Image and Vision Computing Journal*, 23:227–236, 2005.

[Leo96] A. Leonardis and H. Bischof. Dealing with occlusions in the eigenspace approach. In *IEEE Conference on Computer Vision and Pattern Recognition (CVPR)*, pages 453–458, San Francisco, USA, June 1996.

[Li02] C. H. Li and Pong C. Yuen. Tongue image matching using color content. *Pattern Recognition*, 35(2):407–419, February 2002.

[Lip97] S. B. Lippman. *Podstawy jezyka C++*. Wydawnictwo Naukowo-Techniczne, Warsaw, Poland, 1997.

[Loe55] M. M. Loeve. *Probability Theory*. Van Nostrand, Princeton, USA, 1955.

[Lou94] A. K. Louis, P. Maaß, and A. Rieder. *Wavelets*. Teubner, Stuttgart, Germany, 1994.

[Low99] D. G. Lowe. Object recognition from local scale-invariant fearures. In *7. International Conference on Computer Vision (ICCV)*, pages 1150–1157, Corfu, Greece, September 1999.

[Luc81] B. D. Lucas and T. Kanade. An iterative image registration technique with an application to stereo vision. In *Proceedings of DARPA Image Understandig Workshop*, pages 121–130, 1981.

[Ma05] Y. Ma, S. Soatto, J. Koseck, and S. Sastry. *An Invitation to 3-D Vision*. Springer-Verlag, New York, USA, 2005.

[Mal89] S. Mallat. A theory for multiresolution signal decomposition: The wavelet representation. *IEEE Transactions on Pattern Analysis and Machine Intelligence*, 11(7):674–693, July 1989.

[Mal97] Y. Mallet, D. Coomans, J. Kautsky, and O. De Vel. Classification using adaptive wavelets for feature extraction. *Pattern Analysis and Machine Intelligence*, 19(10):1058–1066, Oktober 1997.

[Man02] B. S. Manjunath, P. Salembier, and T. Sikora. *Introduction to MPEG-7 - Multimedia Content Description Interface*. John Willey & Sons Ltd, Chichester, UK, 2002.

[Men99] P. Mendonça and R. Cipolla. A simple technique for self-calibration. In *Proceedings of the IEEE Conference on Computer Vision and Pattern Recognition*, pages 1500–1505, Fort Collins, USA, June 1999. IEEE Computer Society, Washington.

[Mog97] B. Moghaddam and A. Pentland. Probabilistic visual learning for object representation. *PAMI*, 19(7):696–710, Juli 1997.

[Mok88] F. Mokhtarian. Multi-scale description of space curves and three-dimensional objects. In *Proceedings of the Conference on Computer Vision and Pattern Recognition*, pages 298–303, Ann Arbor, USA, June 1988.

[Mor77] H. P. Moravec. Towards automatic visual obstacle avoidance. In *Proceedings of the 5th International Joint Conference on Artificial Intelligence*, Cambridge, USA, 1977.

[Mue68] J. L. Muerle and D. C. Allen. Experimental evaluation of techniques for automatic segmentation of objects in a complex scene. In G. C. Cheng et al., editor, *Pictoral Pattern Recognition*, pages 3–13, Washington, USA, 1968. Thompson.

[Mur95] H. Murase and S. K. Nayar. Visual learning and recognition of 3-d objects from appearance. *International Journal of Computer Vision*, 14(1):5–24, January 1995.

[Nen96a] S. Nene, S. Nayar, and H. Murase. Columbia object image library (coil-100). Technical Report Technical Report CUCS–006–96, Department for Computer Science, Columbia University, 1996.

[Nen96b] S. Nene, S. Nayar, and H. Murase. Columbia object image library (coil-20). Technical Report Technical Report CUCS–005–96, Department for Computer Science, Columbia University, 1996.

[Nga05] Henry Y.T. Ngan, Grantham K.H. Pang, S.P. Yung, and Michael K. Ng. Wavelet based methods on patterned fabric defect detection. *Pattern Recognition*, 38(4):559–576, April 2005.

[Nie83] H. Niemann. *Klassifikation von Mustern.* Springer-Verlag, Berlin, Heidelberg, Germany, 1983.

[Nie90] H. Niemann. *Pattern Analysis and Understanding.* Springer-Verlag, Berlin, Heidelberg, Germany, 1990.

[Pal93] N. R. Pal and S. K. Pal. A review on image segmentation techniques. *Pattern Recognition*, 26(9):1277–1294, September 1993.

[Pap84] A. Papoulis. *Probability, Random Variables, and Stochastic Processes.* McGraw-Hill, New York, USA, 1984.

[Pap00] C. Papgeorgiou and T. Poggio. A trainable system for object detection. *International Journal of Computer Vision*, 38(1):15–33, January 2000.

[Par04] S. Park, J. Lee, and S. Kim. Content-based image classification using a neural network. *Pattern Recognition Letters*, 25(3):287–300, February 2004.

[Par05] Cheong Hee Park and Haesun Park. Fingerprint classification using fast fourier transform and nonlinear discriminant analysis. *Pattern Recognition*, 38(4):495–503, April 2005.

[Pau91] D. Paulus. *Objektorientierte Bildverarbeitung.* Vieweg, Wiesbaden, Germany, 1991.

[Pau92] D. Paulus. Object oriented image segmentation. In *4th International Conference on Image Processing and its Applications*, pages 482–485, Maastrich, Netherlands, 1992. IEEE Computer Society Press.

[Pau03] D. Paulus and J. Hornegger. *Applied Pattern Recognition.* Friedr. Vieweg & Sohn Verlagsgesellschaft GmbH, Braunschweig, Wiesbaden, Germany, 2003.

[Pav82] T. Pavlidis. *Algorithms for Graphics and Image Processing.* Computer Science Press, Rockville, USA, 1982.

[Pol99] M. Pollefeys. *Self-Calibration and Metric 3D Reconstruction from Uncalibrated Image Sequences.* PhD thesis, Faculteit Toegepaste Wetenschappen, Katholieke Universiteit Leuven, 1999.

[Pon87] J. Ponce and M. Brandy. Toward a surface primal sketch. In T. Kanade, editor, *Three-Dimensional Machine Vision*, pages 195–240, Boston, USA, 1987. Kluwer Academic Publishers.

[Pös99] J. Pösl. *Erscheinungsbasierte, statistische Objekterkennung*. Shaker Verlag, Aachen, Germany, 1999.

[Pow81] M. J. D. Powell. *Approximation theory and methods*. Cambridge University Press, Cambridge, UK, 1981.

[Pra01] W. K. Pratt. *Digital Image Processing*. John Wiley & Sons Ltd, New York, USA, 2001.

[Pre90] W. H. Press, B. P. Flannery, S. A. Teukolsky, and W. Vetterling. *Numerical Recipes in C - the Art of Scientific Computation*. Cambridge University Press, New York, USA, 1990.

[Rei01] M. Reinhold, Ch. Drexler, and H. Niemann. Image database for 3-d object recognition. Technical Report LME-TR-2001-02, Informatik 5 (Lehrstuhl für Mustererkennung), Universität Erlangen-Nürnberg, Mai 2001.

[Rei04] M. Reinhold. *Robuste, probabilistische, erscheinungsbasierte Objekterkennung*. Logos Verlag, Berlin, Germany, 2004.

[Rei05] M. Reinhold, M. Grzegorzek, J. Denzler, and H. Niemann. Appearance-based recognition of 3-d objects by cluttered background and occlusions. *Pattern Recognition*, 38(5):739–753, May 2005.

[Req78] A. Requicha and R. Tilove. *Mathematical Foundations of Constructive Solid Geometry: General Topology of Closed Regular Sets*. Production Automation Project, University of Rochester, Rochester, USA, 1978.

[Rip96] B. D. Ripley. *Pattern Recognition and Neural Networks*. Cambridge University Press, Cambridge, UK, 1996.

[Ros71] A. Rosenfeld and M. Thurson. Edge and curve detection for visual scene analysis. *IEEE Trans. Computers*, C-20(5):562–569, May 1971.

[Rue05] H. Rue and L. Held. *Gaussian Markov Random Fields: Theory and Applications*. Chapman & Hall, London, UK, 2005.

[Sch96] B. Schiele and J. Crowley. Object recognition using multidimensional receptive field histograms. In *4th European Conference on Computer Vision*, pages 610–619, Cambridge, UK, April 1996. Springer-Verlag, Heidelberg.

[Sch99] B. Schiele and A. Pentland. Probabilistic object recognition and localization. In *Proceedings of the 7th International Conference on Computer Vision (ICCV)*, pages 177–182, Corfu, Greece, September 1999. ICSP.

[Sch00] C. Schmid, R. Mohr, and Ch. Bauckhage. Evaluation of interest point detectors. *International Journal of Computer Vision*, 37(2):151–172, June 2000.

[Sco00] C. Scott and R. Nowak. A novel hierachical wavelet-based-framework for pattern analysis and synthesis. In *4th IEEE Southwest Symposium on Image Analysis and Interpretation*, pages 242–246, April 2000.

[Sho96] A. Shokoufandeh, I. Marsic, and S. Dickinson. Saliency regions as a basis for object recognition. In *Proceedings of the Third International Workshop on Visual Form*, pages 539–548, Singapore, May 1996. World Scientific.

[Sou95] A. Sourin and A. Pasko. Function representation for sweeping by a moving solid. In *Proceedings of the 3rd ACM Symposium on Solid Modeling and Applications*, pages 383–391, New York, USA, 1995. ACM Press.

[Tad98] R. Tadeusiewicz, W. Wszolek, A. Izworski, and M. Modrzejewski. Application of neural networks in diagnosis of pathological speech. In *Proceedings of the International ICSC/IFAC Symposium on Neural Computation NC'98*, pages 1040–1045, Vienna, Austria, 1998. ICSC Academic Press.

[Tad99] R. Tadeusiewicz. *Introduction to Practice of Application of Neural Networks (in Neuron Networks).* StatSoft, Warsaw, Poland, 1999.

[Tau89] G. Taubin, R. M. Bolle, and D. B. Cooper. Representing and comparing shapes using shape polynomials. In *Proceedings of the Conference on Computer Vision and Pattern Recognition*, pages 510–516, San Diego, USA, June 1989.

[Ter04] D. Terzopoulos, L. Yuencheng, and M. Vasilescu. Model-based and image-based methods for facial image synthesis, analysis and recognition. In *Automatic Face and Gesture Recognition 2004*, pages 3–8, Seoul, Korea, Mai 2004.

[Tom91] C. Tomasi and T. Kanade. Detection and tracking of point features. Technical Report CMU-CS-91-132, Carnegie Mellon University, April 1991.

[Tom92] C. Tomasi and T. Kanade. Shape and motion from image streams under orthography: A factorization method. *International Journal of Computer Vision*, 9(2):137–154, November 1992.

[Tru98] E. Trucco and A. Verri. *Introductory Techniques for 3-D Computer Vision*. Addison-Wesley, Massachusets, USA, 1998.

[Tur91] M. Turk and A. Pentland. Face recognition using eigenfaces. In *Conference on Computer Vision and Pattern Recognition*, pages 586–591, Maui, USA, June 1991.

[Vap95] V. N. Vapnik. *The Nature of Statistical Learning Theory*. Springer-Verlag, New York, USA, 1995.

[Wal99] N. Walsh. *Graphical User Interfaces with Perl: Learning Perl/Tk*. O'Reilly & Associates, Inc., Sebastopol, Canada, 1999.

[Wal00] J. Walter and B. Arnrich. Gabor filters for object localization and robot grasping. In *Proceedings of the 15th International Conference on Pattern Recognition*, pages 124–127, Barcelona, Spain, September 2000. ICSP.

[Web02] A. R. Webb. *Statistical Pattern Recognition*. John Wiley & Sons Ltd, Chichester, UK, 2002.

[Wen93] J. J. Weng, N. Ahuja, and T. S. Huang. Learning recognition and segmentation of 3d objects from 2d images. In *Proceedings of the International Conference on Computer Vision*, pages 121–128, 1993.

[Win94] A. Winzen. *Automatische Erzeugung dreidimensionaler Modelle für Bildanalysensysteme*. PhD thesis, University of Erlangen-Nuremberg, Erlangen, 1994.

[You03] B. You, M. Hwangbo, S. Lee, S. Oh, Y. Kwon, and S. Lim. Development of a home service robot issac. In *Intelligent Robots and Systems 2003*, pages 2630–2635, Las Vegas, USA, October 2003.

[Yua01] C. Yuan and H. Niemann. Neural networks for the recognition and pose estimation of 3-d objects from a single 2-d perspective view. *International Journal of Image and Vision Computing*, 19:585–592, August 2001.

[Zel96] C. Zeller and O. Faugeras. Camera self-calibration from video sequences: the kruppa equations revisited. Technical Report 2793, INRIA Sophia Antipolis, February 1996.

[Zha04] Qinzhi Zhang and Hong Yan. Fingerprint classification based on extraction and analysis of singularities and pseudo ridges. *Pattern Recognition*, 37(11):2233–2243, November 2004.

[Zob03] M. Zobel, J. Denzler, B. Heigl, E. Nöth, D. Paulus, J. Schmidt, and G. Stemmer. Mobsy: Integration of vision and dialogue in service robots. *Machine Vision and Applications*, 14(1):26–34, April 2003.

[Zuc75] S. W. Zucker, A. Rosenfeld, and L. S. Davis. Picture segmentation by texture discrimination. *IEEE Trans. Computers*, C-24(12):1228–1233, December 1975.

Index

Curriculum Vitae

Personal Data

Last Name, First Name	Grzegorzek, Marcin
Date of Birth	28th May 1977
Nationality	Polish and German

Education & Professional Experience

July 2006 – Present	Postdoctoral Research Assistant Multimedia & Vision Research Group Queen Mary, University of London, UK
December 2002 – June 2006	PhD Student and Research Assistant Institute of Pattern Recognition University of Erlangen-Nuremberg Erlangen, Germany
October 1996 – November 2002	Student of Computer Science Institute of Computer Science Silesian University of Technology Gliwice, Poland Degree: mgr inż. ⇔ Dipl.-Ing. ⇔ M. Sc.
June 2002 – November 2002	Programmer Siemens Medical Solutions Erlangen, Germany

May 2002 – November 2002	Master's Thesis Bavarian Research Center for Knowledge-Based Systems Erlangen, Germany
April 2002 – May 2002	Part-Time Assistant Institute of Pattern Recognition University of Erlangen-Nuremberg Erlangen, Germany
April 2001 – September 2001	Sokrates/Erasmus Visiting Student Department of Computer Sciences University of Erlangen-Nuremberg Erlangen, Germany
March 2001	Trainee Programmer KamSoft Company Katowice, Poland
March 2000 – September 2000	Intensive German Course International Graduate School Zittau, Germany
July 2000	Trainee Programmer Mendel University Brno, Czech Republic
September 1992 – June 1996	VIII L.O. im. Marii Sklodowskiej-Curie Secondary School, Katowice, Poland
September 1984 – June 1992	Szkola Podstawowa nr 4 Primary School, Jaśkowice, Poland